Documentation, Disappearance and the Representation of Live Performance

Documentation, Disappearance and the Representation of Live Performance

Matthew Reason

© Matthew Reason 2006

All rights reserved. No reproduction, copy or transmission of this publication may be made without written permission.

No paragraph of this publication may be reproduced, copied or transmitted save with written permission or in accordance with the provisions of the Copyright, Designs and Patents Act 1988, or under the terms of any licence permitting limited copying issued by the Copyright Licensing Agency, 90 Tottenham Court Road, London W1T 4LP.

Any person who does any unauthorised act in relation to this publication may be liable to criminal prosecution and civil claims for damages.

The author has asserted his right to be identified as the author of this work in accordance with the Copyright, Designs and Patents Act 1988.

First published in 2006 by
PALGRAVE MACMILLAN
Houndmills, Basingstoke, Hampshire RG21 6XS and
175 Fifth Avenue, New York, N.Y. 10010
Companies and representatives throughout the world.

PALGRAVE MACMILLAN is the global academic imprint of the Palgrave Macmillan division of St. Martin's Press, LLC and of Palgrave Macmillan Ltd. Macmillan® is a registered trademark in the United States, United Kingdom and other countries. Palgrave is a registered trademark in the European Union and other countries.

ISBN-13: 978–1–4039–9707–4 hardback
ISBN-10: 1–4039–9707–1 hardback

This book is printed on paper suitable for recycling and made from fully managed and sustained forest sources.

A catalogue record for this book is available from the British Library.

Library of Congress Cataloging-in-Publication Data

Reason, Matthew, 1975–
   Documentation, disappearance, and the representation of live performance / Matthew Reason.
     p. cm.
   Includes bibliographical references and index.
   ISBN 1–4039–9707–1
   1. Performing arts – Technique. 2. Performing arts – Information resources. I. Title.

PN1590.T43R43 2006
791—dc22   2006045354

10 9 8 7 6 5 4 3 2 1
15 14 13 12 11 10 09 08 07 06

Printed and bound in Great Britain by
Antony Rowe Ltd, Chippenham and Eastbourne

For Peter and Elizabeth

Contents

List of Illustrations viii

Acknowledgements x

Preface: About the Cover xi

Introduction 1

1 Documentation and Disappearance 8

Part I

2 Archives 31
3 Proper Research, Improper Memory 41
4 Self-Representation 55

Part II

5 Video Documents 73
6 Screen Reworkings 92

Part III

7 Photography, Truth and Revelation 113
8 Photography, Publicity and Representation 146

Part IV

9 Reviewing Performance 183
10 Writing the Live 205

11 The Representation of Live Performance 231

Notes 239

Performance Works Cited 251

Bibliography 253

Index 263

List of Illustrations

7.1	*The Chairs* Complicite and Royal Court Theatre (1997) Photograph Graham Brandon. V&A Images/Theatre Museum	119
7.2	Penny Arcade. *While You Were Out* Performance view at PS122 (1985) Photograph Dona Ann McAdams	124
7.3	Karen Finley. *I'm An Ass Man* Performance view at The Cat Club (1987) Photograph Dona Ann McAdams	125
7.4	Annie Sprinkle. *Post Porn Modernist* Performance view at The Kitchen (1990) Photograph Dona Ann McAdams	129
7.5	*Nachtasiel* Ro Theatre Rotterdam (1998) Photograph Bertien van Manen	131
7.6	Zavarzino Siberia. Oleg (1992) *A Hundred Summers, A Hundred Winters* Photograph Bertien van Manen	131
7.7	*Nachtasiel* Ro Theatre Rotterdam (1998) Photograph Bertien van Manen	133
7.8	Kazan. Vlada (1992) *A Hundred Summers, A Hundred Winters* Photograph Bertien van Manen	133
7.9	Rika Okamoto, Kathy Buccellato and Camille M. Brown Martha Graham Dance Company (1994) Photograph Lois Greenfield	138
7.10	Daniel Ezralow and Ashley Roland (1988) Photograph Lois Greenfield	140
7.11	Wall/Line Ned Maluf, Christopher Batenhorst, Paula Gifford (1994) Photograph Lois Greenfield	142

8.1	Advertising leaflet for *When the Bulbul Stopped Singing* Traverse Theatre Company (2004) Photograph Euan Myles	151
8.2	Advertising leaflet for *Outlying Islands* Traverse Theatre Company (2002) Photograph Euan Myles	153
8.3	'Travel Agency'. Front cover of the Traverse Theatre Brochure (2004) Photograph Euan Myles	156
8.4	Advertising leaflet for *Emanuelle Enchanted* Forced Entertainment (1992) Photograph Hugo Glendinning	159
8.5 a–d	Photographs from the 'blind shoot' of *Club of No Regrets* Forced Entertainment (1993) Photographs Hugo Glendinning	161
8.6	*India Song* Het Zuidelijk Toneel (1998) Photograph Chris Van der Burght	165
8.7	*India Song* Het Zuidelijk Toneel (1998) Photograph Chris Van der Burght	167
8.8	*Visitors Only* Damaged Goods (2003) Photograph Chris Van der Burght	168
8.9	*The Featherstonehaughs Draw on the Sketchbooks of Egon Schiele* The Featherstonehaughs (1999) Photograph Chris Nash	171
8.10	*Blind Faith* Yolande Snaith Theatre Dance (1991) Photograph Chris Nash	173
8.11	*Immaculate Conception* The Featherstonehaughs (1992) Photograph Chris Nash	178

The photographs are reproduced by kind permission of the photographers, companies/artists or rights holders concerned.

Acknowledgements

My sincerest thanks to Graham Brandon, Hugo Glendinning, Lois Greenfield, Dona Ann McAdams, Euan Myles, Chris Nash, Chris Van der Burght and Bertien van Manen for allowing me to reproduce their photographs in this book and for answering questions about their work or giving me the time for interviews.

Thanks also for co-operation with images or other research enquiries to V&A Images/Theatre Museum, Andy Catlin at the Traverse Theatre Edinburgh, Tim Etchells and Forced Entertainment, Jan Coerwinkel at the Ro Theatre Rotterdam, Diane Raspoet at Damaged Goods, Cock Dieleman at Het Zuidelijk Toneel and to Penny Arcade, Karen Finley and Annie Sprinkle.

I am extremely grateful for support for image rights and reproduction from York St John University College and the Society for Theatre Research.

I would also like to thank my partner Alison Dyke, Alice Ferrebe for commenting on sections of the book and Roger Savage for his generous support and encouragement.

Preface: About the Cover

The photograph on the cover of this book is not a documentation of a live performance. It is not a record of something that happened on stage in front of an audience. Instead it is the result of a process begun in a studio and completed on a computer, with the photographer Chris Nash combining four different photographic moments to construct this single final image. This, therefore, is neither a photograph of a performance that actually happened, nor revelatory of something that took place and was visible there in the studio. Instead the form, subject and meaning of this photograph resides only within its own being and existence.

With this photograph, Nash set out to explore two sets of ideas. The first was in terms of movement, interested in developing imagery that was gothic and insect-like in its themes and that emphasised strength, muscle and sinew as well as the dancer's line and form. The second impulse behind the image was in terms of photographic technique, with the intention of getting in close to the dancer with a wide-angle lens and accepting and deliberately utilising the distorting, elongating and foreshortening effects that resulted. In order to control which bits could be elongated and which foreshortened, Nash separated the body up into different sections and then photographed each section doing the particular movement in turn, joining up the different bits afterwards. The forethought and planning involved here is immense, requiring the final image be pre-visualised, taken apart and staged on the dancer's body before being reconstructed within the final work. This process provided Nash with the ability to decide which parts of the body got distorted and to what extent, producing a figure that reads clearly and dynamically and which realises his originating vision.

For the viewer this trickery behind and within the image produces a tension between faith and doubt, between believing the photograph as evocatively real and questioning that it depicts something that could have actually happened. This edge to Nash's photographs provides them with a direct experiential relationship to dance and live performance, even while they remain factually quite removed. For Nash a performance is something that stimulates the imagination, taking place as much in the mind of the audience as it does on the stage, with this active interaction a crucial element to the experience. He hopes to evoke a similar engagement with the viewer through his work, seeking to spark our imagination through interaction with his photographic image.

The photograph on the cover of this book, therefore, is not of a performance and yet is certainly inspirational, evocative and representational of performance. It is also performative in its own right, staging both its subject and the medium of photography. Moreover, it allows us to see and know something of performance that would be unknowable except through the photograph, and in this it provides a useful example of the complex relationship between all representations of performance and their absent and unavailable subject. It is this exploration of how we can know performance through its various representations that forms the theme of this book.

Untitled, September 2005.
Photograph Chris Nash, Dancer Catherine Bennett,
Costume Theo Clinkard, Make-up Louise Nockles.

Introduction

One of the most prominent and recurring definitions of live performance – whether of theatre, performance art, dance or music – is that it is fundamentally ephemeral. More than simply being short-lived or lacking permanency, ephermality describes how performance ceases to be at the same moment as it becomes. Ephemerality describes how performance passes as an audience watches. This is a concept that Peggy Phelan (1993) prominently and influentially articulates in terms of disappearance, providing one of the most potent discourses describing and determining the ontology of live performance. This is a definition, moreover, loaded with significant ideological and artistic weight, as disappearance invests performance with a unique value and radical politics of resistance.

Yet at the same time as it disappears performance also endures. Performance is present and represented in various media and activity that, although not the thing itself, reflect upon, remember, evoke and retain something of performance. This problematic and frequently morally and ideologically loaded concern with the documentation of practice forms a second predominant and preoccupying discourse within performance studies. These contradictory, twin discourses form a key theme in this book, which describes how our cultural knowledge of performance is constructed by the activities, artefacts and representations created within the space between documentation and disappearance.[1]

For Eugenio Barba, theatre is the 'art of the present', with directors, performers, companies and festivals the creators or curators of 'ephemeral works' who must always be focused on the present (1992). Peter Hall similarly describes himself as uninterested in the documentation of his theatre work, which he is happy to see disappear 'like soap bubbles' (2004). Indeed, performance practitioners are frequently focused on the present, attentive to producing and staging new works.

In such descriptions performance only exists in the moment of its creation and its only valid afterlife is in the memory of those who were there. It is different, however, for those who want and need to write and think and teach about performance, whether researchers, academics, critics or students. As Rodrigues Villeneuve recognises, those 'who must speak about the performance' must retain something of it 'something material, some tangible trace' (1990: 32). Here there is indeed both a necessity and a desire to see and know performances that are no longer present themselves.

The result is what Amelia Jones has described as the frequent experiencing of performance as documentation, a phrase used in an article responding to the 'problematic of a person my age doing work on performances I have not seen in person' (1997). By 'my age' Jones was specifically referring to being too young to have seen in the flesh what is now canonical feminist live art of the 1970s and 1980s, but in fact we are all persons of our age, often needing to know and think about work that we have not seen in person. For, given the ephemeral nature of live performance, except by being there in person it is only through its documentations – which in terms of postmodern theory and language we might usefully think of as resonances, traces and fragmentations – that it is possible to know, question or see performance at all.[2] Additionally, for large components of their study, this is often *the* experience of the university theatre student. Performance studies courses are in the business of teaching students about performances of the past and of the present. Often, however, what they are actually getting students to do is look at representations and documentations of performance – videos, photographs, archival documents, reviews, other writings and so forth. While such problems are not necessarily ignored they are underplayed within performance studies as a discipline; the explicit intention of this book is to refocus attention on what it means to see performance through its documentations.

Here there are a series of crucial questions, with Jones herself asking whether being *there* in person provides access to a superior kind of truth about performance. Similarly, we need to think about the exact relationship between seeing a documentation and seeing a performance, about what kind of knowledge of performance we can access through its representations, about the interpretations present within each act of representation and about the tension existing between documentation and the positive valuation of performative disappearance. In exploring such relationships we must consider whether we are thinking in terms of qualities of authenticity, accuracy, completeness and reliability (all particularly relevant in terms of historical research and knowledge), in terms of the evocativeness or beauty of the representation in its own right, or alternatively about emotional, artistic or social truths and appropriateness. All these questions are particularly relevant to those engaged in the evolving area of practice-based research. Here there is both an investment in the undissolvable experience of doing and being (particularly from the point of view of the performer). Yet at the same time there is a need to extract and externalise this knowledge, particularly in the form of documentations that can endure, be more widely disseminated and known by those who were not there.[3]

Inspired and motivated by this range and complexity of questions, this book is concerned with thinking about what exists outside or beyond or after live performance itself – things surrounding, about, reflecting upon and re-presenting performance in one form or another. Things that are not performance, but which allow us to see and say something about performance. Moreover, the interest is not in documentations as passive and transparent windows on to performance, nor in condemning them as something problematic to be overcome or begrudged, but instead in using them as an interrogative opportunity by which we may interpret performance. The interest, in short, is in how these representations make performance knowable.

The use of the word 'representation' in this context needs a little definition. Following the event itself, fragments and echoes of the performance are re–presented in various media and activities, with sections of this book focusing on video recordings, still photography and writing amongst other possibilities. The enduring, though modified, existence of such traces of performance is one aspect of this use of the word represent. There are, of course, obvious differences between, for example, the manifestly (although not absolute) mimetic nature of a video recording and the linguistic evocation of a performance in writing – and the specificities of each medium are central to the analysis of the representations. Nonetheless, in this context each has a similar relationship to its subject: each allow performance to be known, discussed and seen (in a modified 'representational' form) beyond the moment of its creation; each forms a cultural field that represents performance for particular social purposes. Additionally, this interest in representations of performance is directed precisely by the limitations, failures and omissions resulting from the fact that they are not absolute reproductions of the thing itself.[4]

For importantly, these representations of performance do more than merely seek to halt disappearance. Instead, with all representations, the methods and interests of the presentation also begin to constitute a distinct identity of its subject. That is, in the choices of *what* to record, in the manner of *how* to record and indeed in what *can* be recorded, the act of representation defines its subject. Consequently, the transforming and mutating impact of each representation provides an opportunity or fulcrum through which to explore the cultural perceptions, valuations and understandings of the nature of the performance represented.

This book, therefore, explores the theoretical, conceptual and even political and moral implications of how the live arts – particularly theatre and dance – are recorded, described, documented and otherwise represented beyond the moment of their manifestation in performance. It uses a verbal and visual discursive and representational analysis to explore how these enduring afterlives and afterleavings of performance construct what performance is as a cultural phenomenon. The intention is not to present a survey or catalogue of performance representations and documentations, but explore, adopt and develop approaches and interpretative strategies that allow us to use these representations as an insightful and self-questioning form of knowledge and way of seeing. Focusing on contemporary performance practice, and through conceptual discussions, close analysis of case illustrations and interviews with practitioners, this book explores how we can know live performance through its representational traces.

Chapter summaries

Chapter 1 presents the theoretical, political and possibly moral implications of the tension between documentation and disappearance. This chapter identifies how the space between documentation and performance makes disappearance manifest and visible. Moreover, it is this space between documentation and disappearance, a space of fragmentations and representations, which forms the site of our cultural knowledge and vision of performance.

This conceptual framework provides the grounding for the subsequent chapters, each exploring the transformational and perceptual impact of different media, discourses or activities of performance representation. With theatre and dance of necessity often known through their various representations, it is vital to place the analysis of attempts to document, record, remember and thereby represent live performance at the centre of critical considerations of the performing arts. In sections divided according to medium or activity, each chapter in this book explores both the theory and practice of representing performance. Considering the translation of ephemeral performance to lasting document, the chapters examine what impressions the representations leave on our understanding of performance and in turn how the nature of performance determines how we respond to and interpret the representations.

Part I on performance, archives and the academy presents three activities of documentation and representation that are able to claim, for differing reasons, a powerful degree of authority and legitimacy. **Chapter 2** explores how the archive in particular is often constructed as the most 'proper' storehouse of performance afterlife. This discussion examines the very concept, promise and practice of the live performance archive, uncovering the metaphors than run through archival discourses, particularly those concerning memory and retention. Following this exploration of archive theory, **Chapter 3** examines the relationship between the academy, research and the afterlife of performance. Again this is an activity of documentation possessing an aura of authority, this time in part as a result of its production by 'experts'. This chapter considers the tension between the values and claims of 'proper research' in the academy and the positive perception of memory as an alternative site of performance afterlife. As a case study investigation bringing together the issues explored in these two chapters, **Chapter 4** looks at the documentary drive that is apparent in the work of Forced Entertainment, describing the relationship between archival and academic discourses of documentation and the company's own acts of self-representation. This chapter considers how, although now in a very different manner, such self-representations also possess a documentary authority – this time as a result of their authorship by the originating artist.

The following three sections explore what are the most prominent, influential and widely circulated forms of performance representation, considering in turn video, still photography and reviews.

Part II presents two chapters exploring the relationship between audio/visual recording technologies and live performance. Describing it as an activity or set of attitudes as much as a particular technology, **Chapter 5** examines how video is often presented as both the potential saviour of theatre and dance (providing documentations of the performance in time and space and thereby halting disappearance) and as a potential threat (the lasting video representation usurping the absent live performance). This chapter explores the significance of this tension, identifying how it impacts on attempts to document live performance in a non-live medium. **Chapter 6** widens discussion of audio/visual recording to consider more diverse approaches to the reworking of performance on screen, exploring how such hybrid forms impact upon how we watch and understand both the screen and the stage.

These two chapters present conceptual considerations of the relationship between video, technological recordings and live performance, employing case study illustrations of the work of a number of companies and practitioners to review the practices and impulses that exist in this field. Through this analysis, the discussion reveals how the discourses surrounding the video representation of theatre and dance construct a cultural perception and valuation of the particularity of liveness as a medium of performance.

Part III explores the enduring importance of still photography, which remains the most frequently used and seen representation of performance. Over two chapters the contrasting and competing revelatory and representational aesthetics of photography are examined, paying particular attention to the circumstances of production and narratives of use behind performance photography. **Chapter 7** considers the enduring and powerful articulation of photography as a revelatory medium, able to isolate and reproduce a moment of truth, and the tension this promise produces when seen in relation to the dynamic form of live performance. **Chapter 8** considers photography as a more interventionist and painterly medium, particularly in the context of its use as a marketing and publicity tool. This chapter also considers the impact of digital technology on photographic discourse and the common spatial and chronological alienation of performance photography from actual live performances. Each chapter explores the work of a number of photographers, between them demonstrating a range of different styles and approaches to performance photography, discussing how the choices enacted by the frame of the camera presents a selective, value laden and interpretative representation of its subject. The contrasting practical, ideological and representational implications of these approaches are explored in detail.

Part IV presents an extended discussion of writing about performance, particularly in the context of newspaper reviews. Consideration of the review has often, and importantly, looked at the social and artistic power of the critic. While keeping this vital context in mind, these two chapters consider writing as a representational act, looking at reviews as one of the most prominent and immediate instances where language is required to account for performance. **Chapter 9** examines what aspects of live performance events are presented in reviews and what omitted, exploring how the form and content of reviews is directed by their various functions and circumstances of production. **Chapter 10** broadens this analysis, exploring the representational qualities of writing and developing some more speculative imaginings about potential representations of live performance in language. In particular, this chapter engages with concepts of phenomenology and tests the possibility of encoding liveness into the very language and structure of the review. Discussion in both chapters focuses first on a series of intense and detailed readings of how the language of reviews describes and constructs the production for the reader.

Across these chapters, this book explores a range of technological, linguistic and archival activities of representation existing within contemporary performance practice. In very different manners each of these activities aims to account for, remember and thereby represent the ephemeral performance. Operating in fundamentally different media than their subject, each of these activities of representation must perform choices and transformations – with every selection and omission constructing and communicating particular conceptions of the performance represented. This book uses such transformative acts of documentation to explore the cultural meanings and perceptions placed upon performance and reveal the representational implications of the theory, politics and practice of documenting live art.

As a conclusion, **Chapter 11** develops and reviews the rationale for this focus on the afterlife of live performance, arguing that representations of performance are partly a matter of what can be communicated – given the nature and limitations of the representing medium – and partly a matter of choice, objectives and the cultural valuation of the subject. All documents and traces of live performance, therefore, must be considered as presenting cultural, political and artistic perceptions of their subject and at the same time must be analysed with an understanding of the abilities, traditions and objectives of the representing media. The result is a methodology of performance analysis that focuses our attention on how we might employ representations of theatre and dance as an active, interrogative tool of knowing performance.

1
Documentation and Disappearance

The cultural meaning of performance resides not only within the thing itself, but is also and significantly constituted through attempts to describe and define what performance is as a phenomenon. It is with this perspective that this chapter explores two prominent and ongoing conversations about performance: the first of disappearance and transience; the second of documentation and retention. These discourses occur primarily within the academy and performance studies, but also filter out into more generalised dialogues within the cultural industry, amongst audiences and in the arts media.

Discussion starts with an exploration of the prominence of ideas of ephemerality and liveness within articulations of the ontology, ideology and experience of performance. Here there exists a positive valuation of live performance as something that disappears, particularly when constructed in opposition to the recorded and re-playable media of non-live performance. At the same time, however, there also exists a parallel discourse articulating a strong cultural fear of transience and a resulting desire to document performance and thereby save it from disappearance. Together these two themes form a contradictory, mirroring discourse of documentation and disappearance and this chapter explores how, both separately and together, these discourses construct their subject, particularly through the powerful tension that exists between them. Indeed, this chapter argues that it is the paradoxical instincts and values contained within these twin discourses that are central to cultural constructions of performance.

Transience

Containing within it all the implications of liveness, one of the concepts most frequently evoked in descriptions of performance, in whatever form or genre, is that of transience. That performance passes in time; that it has no physical durability or permanence; that its existence is temporary, momentary, fleeting is a centrally repeated definition of theatre, dance, music and the live arts generally.[1] This discourse of transience constructs live performance as something that not only exists in time, but which exists in a significant, special, performative and ephemeral time.

Awareness of the centrality of this discourse of transience exists first in each of our own experiential knowledge, whether as audience member, researcher or practitioner. We can, similarly, call upon anecdotal evidence present in the sheer number of times that performance is described as transient, ephemeral, fleeting, temporary, momentary – the words re-doubling, multiplying and returning on themselves as they are used interchangeably and reinforcingly. For the language and idea of transience is a central motif of many ongoing conversations of and about live performance.

The idea of transience is particularly prominent, however, in discourses produced by those most closely involved in the arts sector – actors, directors, practitioners – amongst those with a personal or social investment in theatre, dance or performance as art forms – self-declared aficionados, audiences defined as 'frequent attenders' – and, not least, amongst those involved in writing or thinking about performance, whether theorists or critics. Indeed, as Rebecca Schneider writes, the perception of 'performance as an "ephemeral event" has been a cornerstone of Performance Studies, and has been evident as basic to performance theory since the 1960s' (2001: 106). Additionally, Gay McAuley not only stresses perceptions of transience when she observes that 'Theatre, by its nature, is an art of the present moment', but also continues to suggest that 'it is fascinating to performers and audiences precisely because it is unique and ephemeral' (1994: 184). Transience, in other words, becomes not merely one phenomenological characteristic of live performance, but also a motivating and inspirational virtue. That this discourse is strongest amongst those with an investment in (or sense of cultural ownership of) performance is significant, immediately suggesting that it possesses a particular centrality and importance.

Accepting the familiarity of the concept, it is nonetheless valuable to elaborate on what exactly is meant by a discourse of transience, starting with direct references and descriptive statements where performance is defined as ephemeral. Examples of this can be drawn from a multitude of sources within a vast discourse, whether it be Michael Kirby observing how 'performance is perishable' (1974b: 1), Marcia Siegel writing of how 'Dance exists at a perpetual vanishing point. At the moment of its creation it is gone' (1972: 1), Laurie Anderson declaring that 'Live art is especially ephemeral' (in Goldberg, 1998: 6) or Eugenio Barba stating, ontologically, that 'The theatre's nature is ephemeral' (1990: 96). Taken collectively these individual statements reinforce and reiterate each other to form a discourse that defines its subject in its very tangle of interconnections. Recognising this, a feature of much of the following discussion is the presentation of a myriad of voices and public conversations, used as self-demonstrating illustration of the re-doubling nature of the cultural perceptions of performance that they construct. Additionally, these statements of the transience quickly become more than merely descriptive, but also declarative – not merely observations, but artistically and ideologically loaded manifestos.

Manifestos of nowness

As a concept, transience is essentially concerned with the business of time; specifically with the location of performance within time. The transient is that which passes away with time, being temporary, brief, momentary, fleeting. Dictionary definitions link transience to the ephemerality of flowers, insects and diseases: as life 'beginning and ending in a day'. Transience is that which is perishable – containing connotations of death and decay, which will become increasing relevant through the course of this chapter – with its existence determined by its location in time. Transience, therefore, describes how performance exists in some kind of unique time: described as unfolding in its own temporal frame. Indeed, time and ephemerality are inherently bound together, as when Merce Cunningham writes that dance gives nothing back but a 'single fleeting moment' (1968); or Thornton Wilder declares that, 'The supremacy of the theatre derives from the fact that it is always "now" on the stage' (in Cowley, 1962: 100). This 'fleeting moment' or 'now' provides a succinct and revealing description of the unique temporal locatedness of the experience and existence of live performance. Indeed time and ephemerality can only come together in the present moment, in what can be described as the 'performative now'.

The overriding implication of the *now* of performance is its statement of presentness, something Bernard Beckerman reveals in his description of theatre as a temporally determined form: 'Theater is nothing if not spontaneous. It occurs. It happens. The novel can be put away, taken up, reread. Not theater. It keeps slipping between one's fingers' (1979: 129). The imagery employed here is significant (although not unproblematic, as will be returned to later), with 'spontaneous' hinting at no prior existence and 'slipping between one's fingers' at no subsequent afterlife – existence, in other words, only in the *now* of performance.

And it is possible to find numerous similar formulations of this existence of performance only within the present now. Such articulations include Peter Brook, who writes that a performance 'is an event for that moment in time, for that [audience] in that place – and it's gone. Gone without a trace' (in Melzer 1995a: 148), Susanne Langer, who provides a description of music in which ' "now" turns into unalterable fact' (1953: 139) and Josette Féral, who declares that 'Performance escapes all illusion and representation. With neither past nor future, performance *takes place*' (1982: 177). Or, returning to what may be one of the primary influences on much of this discourse, Antonin Artaud, who describes theatre as 'the only place in the world where a gesture, once made, can never be made the same way twice' (1958: 75). Declarations such as these maintain that there is no possibility of repeating performance, that it can have no independent life outside of the event itself and no undistorted existence beyond the state of being that is the present moment of its creation.

The similarities and reiterations within these statements form a powerfully constituting discourse that seeks to define the cultural meaning of performance. Indeed, such is the prominence of this discourse that direct statements asserting that one or another form of live performance is transient are simply too frequent to count and too repetitive to reproduce any further here. It is this multiplication and replication that marks assertions of transience as a central discourse of performance. Indeed, the value and importance placed upon such discourses is evidenced by how declarations of ephemerality frequently position different forms of live performance as if in open competition, each vying to assert that one is the more ephemeral than the other. That this appears to be an active competition indicates how transience becomes, discursively, more than a description but an accolade; and more than something desirable but also a manifesto of intent and primary purpose. In other words, the transience of performance becomes a positive and indeed essential description of the thing itself. Further, transience is constituted by these discourses as central to the unique and genuine experience of performance, with it often implied that the 'greater' the transience the more valuable the experience. Ultimately transience becomes a description not only of the ontology of the live performance, but also an aesthetic value in its own right and a political statement of the ideology of performance.

Disappearance

If live performance is to be defined its transience, then it must also be defined by its disappearance: as Patrice Pavis writes, 'the work, once performed, disappears forever' (1992: 67), Herbert Blau similarly articulates performance as 'always at the vanishing point' (1982: 28), while Adrian Heathfield describes how performance exists at the 'instance of disappearance' (2000: 106). Ideas of disappearance, absence and memory are the inevitable product of transience, with such imagery not only describing the nature of the experience but also coming to constitute an ideology and cultural meaning of liveness. Bringing these two points together, discourses constitute performance as existing in circumscribed time, present only for the moment and then gone. What is valued as 'performance' is valued for the transient moment of its creation and for its subsequent, inherent disappearance.

The ontological and ideological import of disappearance is provided its most influential articulation by Peggy Phelan in *Unmarked: The Politics of Performance* (1993). Here Phelan locates performance in qualities of 'disappearance', implicitly bringing together many recurring discursive motifs of transience and liveness. In a passage that has become the classic formulation of this position, Phelan writes that 'Performance's only life is in the present', continuing:

> Performance cannot be saved, recorded, documented, or otherwise participate in the circulation of representations *of* representations: once it does so, it becomes something other than performance. To the degree that performance attempts to enter the economy of reproduction it betrays and lessens the promise of its own ontology. Performance's being ... becomes itself through disappearance. (1993: 146)

Existing within established discourses of transience and manifestos of the performative now, over the last decade Phelan's articulation of disappearance has become one of the most prominent and frequently repeated expressions of performance culture. It has been adopted and repeated far beyond the bounds of the specific genres of live art and the context of psychoanalysis and identity in which it was originally formulated. This is not surprising, as Phelan presents an attractive statement of performance as possessing an innate ideological strength, with a unique worth, purpose and even moral value residing in its disappearance.

What quickly becomes a central difficulty with Phelan's argument, however, is that this ontological and ideological assertion of disappearance also rests upon a practical and empirical statement of non-reproducibility. As Phelan notes, while largely aiming to comment on performance in relation to commodity exchange:

> The response to *Unmarked* has been much more about the technology of the ideal performance archive and the nature of disappearance. ... I was not saying, although I've heard people say I *was* saying, that we must not have photographs, videos or sound documentations of performance. I'm quite happy to have those. ... If I can paraphrase myself reasonably successfully, 'performance betrays its ontology to the degree to which is participates in the economy of reproduction.' That's not exactly it, but it's close. This word 'betrays' has been a bit of a problem, I think I was read as a high priest saying 'we must not have betrayal!' (2003: 294–5)

Phelan is self-conscious about how her words have become prominently adopted within discourses of performance, ripe for paraphrase as they assume a life and significance far beyond their author's control. Indeed, the impact of her statement has to a great extent been in response to its discursive potency and succinct formulation – performance becomes itself through disappearance – of what was already a prominent and recurring valuation of transience. Nonetheless, despite the runaway discursive afterlife of her language (and the slight mis-reading of her argument) it is clear that for Phelan, and for others, disappearance quickly becomes more than a description, and more even than an ontology, but also a moral, political and attitudinal statement and assertion of the importance and value of performance.

The politics of disappearance

Phelan's identification of disappearance as the key ontological and political characteristic of performance has been widely adopted, including by Adrian Heathfield and Andrew Quick in a *Performance Research* editorial:

> as Peggy Phelan has argued, the distinguishing (and radical) feature of performance is its very ephemerality; in its disappearance it evades the knowing and commodifying clutch of representation. (Heathfield and Quick, 2000: 1)

This radical, ideological importance of disappearance is something also identified by RoseLee Goldberg, who describes how live art culture ascribes positive value to the intangible nature of performance, its resistance to being bought and sold and the way it leaves no traces behind (1988: 152). Catherine Elwers similarly writes that performance art held positive the fact that 'no object remained after the event to be collected, sanctioned and sanctified by the critics, historians and collectors controlling the arts establishment'. She continues, however, to note that 'the cannier 1960s artists carefully kept the detritus of their live work whilst protesting their leftist disinterest in objects of performance, which are now increasingly collectable' (2005: 6). The serious implication behind this passing observation is the extent to which live art can be exchanged and consumed as a product, and to what extent it is politically resistant to commodity exchange or does in fact circulate within systems of reproduction and representation.

Indeed, this tension within the very concept of performance as transient, as disappearing and as thereby ideological and ontologically different from non-live performances and material culture, has recently become the central motif in an important and ongoing dialogue within performance studies. This has been particularly in terms of Philip Auslander's book *Liveness: Performance in a Mediatized Culture* (1999), which directly responds to and critiques Phelan's ontology of performance.[2] Auslander's central argument is that ontological conceptualisations of live performance have escaped for too long without serious critical examination, becoming unchallenged through their repetition and reiteration within discourses of performance studies. In contrast he seeks to highlight how the status, perception and even definition of live performance are culturally contingent and neither ontologically nor technologically given. His critique is also significant in how it positions the debate within key postmodern concerns, including the systematic questioning of essentialism, the highlighting of the position of culture within systems of reproduction and exchange and the deconstruction of ideas of origin and authenticity. In this context this debate provides a useful tool to further focus attention on the issue of what the disappearance of performance really signifies.

In *Liveness* Auslander questions the grounds on which distinctions between live and (to borrow his terminology) 'mediatized' performances are constructed, highlighting the implicit prejudices and unconsidered value judgements frequently involved in the privileging of the live over the non-live. As a central component of this critique Auslander directly challenges Phelan's concept of performance becoming itself through disappearance, no doubt seeing her work as one of the most influential articulations of performance as ideologically radical and ontologically distinct from mediatized performances. In response Auslander argues that not only does performance systematically repeat and reprise itself, but also that it is always already an act of reproduction or representation.

Auslander not only rejects the validity of *any* ontological definition of live performance, but also seeks to blur and overturn any practical and experiential distinctions between the live and the mediatized – particularly in terms of transience and disappearance. Taking popular music, theatre and television as his principal territories, Auslander works through a lengthy series of investigations displaying the mutual entanglement of the live and non-live, determinedly working to question, erode and finally discard as irrelevant, insignificant or unconvincing any residual differentiatations. In each example he presents, Auslander asserts that live performance is always already embedded within systems and industry structures of production and re-production. Here matters of cultural practice and performance production challenge the articulation of performance as not only commodity resistant but also as essentially defined by disappearance. For on a practical level the idea of the 'presentness' and 'unrepeatability' of live performance is shot through with paradoxes and contradictions.

Performance as commodity

Not least amongst the problems raised by the concept of transience and disappearance is how we account for the clearly apparent existence of 'production runs' and 'repeat performances', each referring backwards to previous occurrences or forwards to future existences. Such repeatability is built-in through economic necessity, yet nonetheless within the culture of the live arts each individual performance continues to be articulated as essentially unique and ephemeral, with transience thereby paradoxically repeated each night of a repeated performance. To an extent subtle differences do distinguish one performance of the same production from another. Particularly to those involved in the minutiae of a production, each night of a run of the same performance is different: differences caused by innumerable variations, both small and large, conscious and unconscious, and differences caused by variations in an audience's responses and reactions. The 'unchanging' script or score aside, a particular live event is immediately perceived as constructed from a combination of that particular performance and that particular audience, a combination that cannot be repeated. As such, performance does indeed seem to resist description either as something that is consumed[3] or as something that is mass-produced as a commodity.

Such arguments are legitimate, presenting a compelling depiction of the uniqueness and unrepeatabilty of live performance. However, the idea that every performance of a production is different can also be seen as less a concrete description than a partially accepted and discursively constructed convention. For any distinctions are usually (although not always) minor and not of such an extent to warrant a repeat performance being termed and seen as a distinct event in its own right. The extreme illustration of this occurs with the celebration of the reproductive milestones of long-running productions, such as the marking of the 20,000th performance of *The Mousetrap* in London on 16 December 2000. Although each of these performances might be perceived as different, the very fact that a 20,000th performance can exist at all manifests an demonstrable sameness and suggests that live performance is evidently repeatable.[4]

More significant than the possibility of tangible and observable (if minor) differences in a production, is our knowledge of the fact of reproduction. This meets its ultimate manifestation in commercial productions, such as the Broadway and West End musical or the Disney stage show. The franchised, touring, stage version of Disney's *Beauty and the Beast*, for instance, is effectively the live reproduction of a film 'original', replicated simultaneously in venues across the world. This live Disney production is promoted with the slogan 'The Magic Comes Alive On Stage', a deliberate packaging and promoting of the live experience that is both an invitation to the temporal ('comes alive') and spatial ('on stage') 'magic' of live theatre and yet at the same time reminds us that it is a copy of a non-live original. Certainly a great deal of the motivation to see this live reproduction is created by the existence of the media version. Here, rather than radically resistant, performance becomes intrinsically embedded within cultures of commodity exchange and mass reproduction. Indeed, liveness itself becomes the commodity in question.

Among others exploring this theme, Dan Rebellato describes how the production processes of the mega-musical shares with global capitalism the key components of automation and franchising. This he notes has earned it 'the unkind but apposite nickname of McTheatre':

> A founding principle of McDonald's was that every Big Mac, wherever you were in the world, would taste the same. Cameron Mackintosh similarly – and commendably – insisted that his shows should not become any less professional and polished the further in time or space they were from the first press night. However, as the production process becomes more and more automated, what begins

as a guarantee of quality ends as a guarantee of predictability. As his biographers write, intending, I think, to be flattering, '*Cats* was effectively and expensively reproduced around the world as exactingly as any can of Coca-Cola and wherever you saw it, the sensation was the same'. (2005: 102)

Here performance becomes a commodity that is repeated as exactly as possible not just night after night but year after year regardless of cast changes, the employment of new directors or the restaging in new venues. 'These shows are commodities', writes Rebellato, 'and the attitudes behind them are nakedly about capital accumulation'. In such circumstances the show, set and brand are more important than the actor, who is reduced to replicating not just what was prepared in rehearsal but also what might have been originally conceived for a different performer several years previously. This is something Susan Russell observes when drawing on her own experience as a musical performer:

> I was one of thirty-seven workers who built the standardized product of *The Phantom of the Opera* every night. My function was to replace a missing worker, accomplish their required tasks, and assemble the product without missing a beat, interrupting the flow, or disturbing the rest of the machine. (2003: 57)

Russell describes this experience as one of 'forced replication', with the creativity of the artist exchanged for an assembly line 'corporate actor'. This evident commodification of liveness in the commercial realm causes problems for descriptions of the live as a site of political and ideological resistance. Indeed, here the very being and presence of the actor becomes commodified and systematically reproduced.

Accompanying descriptions of the replicated live performance, it is also worth remembering the familiar stories of how some individuals attend the 'same' performances on dozens and even hundreds of occasions. In other words the experience itself seems to have become repeatable. The explanation for this is in part a matter of the successful marketing of liveness as a product, as Russell writes 'the "now" is what corporate producers want' (2003: 59). And according to the terms of such marketing, while promoting a unique, one-off, live experience it is likely that such repeated consumption is the result of the standardisation of the product (its quality of already being known) rather than of any truly radical experience or expectation of difference.

It is important here to acknowledge that such commercial theatre is not necessarily 'performance' in the manner that Phelan and others have primarily in mind when discussing the radical resistance offered by disappearance. However, while not conflating all distinctions entirely, it is valuable to juxtapose commercial theatre with live art in order to examine the radical and ideological claims that some commentators invest in liveness. For while live art and other marginal performances exist within fairly coherent systems of production and reproduction, *disappearance* in this context is not solely concerned with transience but also seeks to assert the status of the experience as not merely unrepeatable but also as unrepresentable – as unspeakable experience, as sublime experience. A stark contrast, in other words, to the sameness or knownness suggested as inherent to the experience of commercial theatre. While commercial performance may involve forced replication, live art asserts an absolute presence and realness whereby performance and performer are inseparably rooted within the physical body.

However, it is noticeable that in terms of *discourse*, both commercial theatre and more marginal live art extol many of the same values: presence, uniqueness, difference, transience. Additionally, the perhaps more difficult task of asserting the disappearance of live performance even when manifested in commercial theatre presents the more interesting and important possibilities in this context – and one that is valid to constructions of the experience of all live performance. For example, in exploring precisely the production runs in the commercial theatre described here – where 'when an actor replaces the original performer of a role he is expected to follow the format of his predecessor, a format that success has sanctified' – Bernard Beckerman invites us to consider an alternative possibly, found in a more elusive understanding of the performative 'now' as a *promise*, a contract of presentness made between performance and audience, which a long run, series or tour seeks to reiterate artfully on each occasion. Liveness, as distinct from life, is not a matter of only happening once, something that Beckerman describes eloquently when he writes that the promise of live performance is to reiterate and renew the now:

> even when the essential shape of a scene is repeated, the manner in which it is filled by the spontaneous energies of the performer often produces a significantly distinctive experience. All the patient care of rehearsal and planning is for the purpose of presenting a spontaneous moment, a moment that is unique to that company and that audience at a particular time. (1979: 161)

Beckerman's use of the word 'spontaneous' in terms of concepts of transience has already been commented upon. Here it is again striking, yet must prompt a moment of hesitation in the face of our awareness of the intricate levels of rehearsal and reproduction involved in such performances. However, literal spontaneity is in many ways a prosaic understanding of the performative now. More interestingly it is the *idea* of the now that Beckerman suggests is reconstituted every night. Performance promises to be the present, promises to be unrepeatable presentation. This is what performance *promises* even when it is recognised that it will not be fulfilled literally. For audiences, however, the promise of presentness, and acceptance of that promise, is more important than absolute temporal uniqueness. In a willing contract between the audience and the performance this promise of presence is not a delusion, but instead an event that creates faith; it is a promise that also carries with it its own reward. The mass replication of this promise through marketing and in commercial theatre does not necessarily reduce its power and experiential impact.

Somewhat analogously, Walter Benjamin argues that manual reproduction enhances the idea of authenticity, suggesting that the labour of reproducing a work by hand maintains the unique quality of both the original and the copy (1970: 214). Although severely challenged by the description of the actor as automaton provided by Russell, it nonetheless is the case that the reproduction of live performance is dependent, as Jonathan Miller suggests, on a process of manual reproduction (1986: 67). And of course, Phelan never argued that performance cannot be repeated, instead more subtly writing that 'Performance occurs over a time which will not be repeated. It can be performed again, but this repetition marks it as "different" ' (1993: 146).[5]

These questions as to the nature of repetition, and the relationship between repetition and 'generality' or reproduction, are developed in depth by Gilles Deleuze in *Difference and Repetition* (1994). Here it is worth briefly pointing towards how Deleuze writes in his introduction that

> To repeat is to behave in a certain manner, but in relation to something unique or singular which has no equal or equivalent. And perhaps this repetition at the level of external conduct echoes, for its own part, a more secret vibration which animates it, a more profound, internal repetition within the singular. This is the apparent paradox of festivals: they repeat an 'unrepeatable'. They do not add a second and a third time to the first, but carry the first time to the 'nth' power. (1994: 1)

Each repetition, therefore, is different. Each repetition is singular and distinct by the very nature of it being a repetition. In this way repetition is itself a key marker of difference and in many ways the instigator or creator of difference in the first place.

The difference between repetition and reproduction is crucial, something that can be seen in the relationship between live and non-live media of performance. Although it is largely neither useful or necessarily to pursue hard and fast, ontological differentiations, an appropriate and meaningful distinction does exist in the way that live events are re-performed while non-live performances are re-played. This distinction in the manner of repetition is crucial, highlighting how the dynamic process of re-performance continues with each live performance (and is manifested in the very repetition), but has ended and ceased to be dynamic with non-live media which are not themselves repeated but instead remain the same on each replaying.

The transience of performance, therefore, is visible not only in the manner by which it disappears, but also in how it endures – in how it is repeated, re-presented, re-performed. Rather than being dependent upon the lack of reproduction, transience is visible in the effort that goes into eradicating change and staging the repeat performance. This effort, even when pushed to the extent and oppressiveness of the megamusical, marks live repetition as different from mechanical reproduction. Similarly, disappearance is the result of (and always accompanied by) a sense of retention. Just as knowledge of loss is only possible through the act of memory, so equally (and to paraphrase Phelan) does disappearance become possible through documentation.

Becoming through documentation

Peggy Phelan's statement that performance 'becomes itself through disappearance' presents transience as an ontological definition (and also an ideological value) of performance. Meanwhile remarks by Jean Genet on the future life of a theatre production illustrate the acceptance and appreciation of disappearance in practice:

> It will not be possible for all the living, the dead, and the future generations to see *Les Paravents* ... All the stage performances which will follow the first five ones will only be mere reflections. That's what I think will happen. But anyway, who cares? One well-rehearsed performance should be enough. (In Kalisz, 1988: 79)

The positive value placed here on the limited lifespan of performance celebrates ephemerality. As Genet asks, why should anything more be necessary? Similarly, Eugenio Barba declares theatre the 'art of the present' and describes directors and performers as creators of 'ephemeral works' who must always be focused on the present (1992: 77). Artists, like Barba and Genet, who celebrate the transitory nature of their work often repeat this primary valuation of transience. George Balanchine, for example, was (according to his associate Barbara Horgan) 'a man who didn't give a damn about the past and cared even less about the future', instead committed to working in the present moment of new performances (in Brooks, 2001).

However, while Balanchine may not have been interested in the future life of his work, Bonnie Brooks observes with evident relief that others certainly were: 'An examination of practices in the dance field shows that the work of saving dances often appears to fall not to the artists themselves, but to the people who surround them' (2001). The work of 'saving' dances is one of saving them from disappearance and these comments and documentary ambitions are repeated across the performing arts. As here, the expressions of documentary need are often unreflective; the obvious question as to why we might desire retention is unasked. This is because the answer is always already implicit: if we do not document performance it disappears; we document performance to stop it disappearing. Already, this represents a significant ideological step away from Genet's belief that one performance should be enough and far away from any positive valuation of disappearance. In discourses of performance, disappearance and transience mark one set of recurring imagery, but they are accompanied by a mirroring, complementary and contradictory discourse of documentation.[6]

Evil of oblivion

The desire to document performance is a strong, contradictory thread running through the live arts. It is a desire motivated by the same perception of transience explored earlier, yet now seen as something negative rather than a cause for celebration. Indeed, it is the fact that performance is live that motivates this instinct to documentation. As Ben Jonson writes in his preface to the publication of *The Masque of Blackness* in 1608, his script exists because the performance could not last:

> The honor and splendour of these spectacles was such in the performance as, could these hours have lasted, this of mine now had been an unprofitable work. (Jonson, 1969: 47)

The underlying motivation for the documentation of live performance, therefore, is always the same: unable to hold performance continually in the present, it must be translated into some more enduring if less splendid form – it must be represented. And Jonson's preface continues, adding that his work in publication intends to 'redeem' his work in performance from the common 'evil' of 'oblivion'.

This fear that performance disappears if active steps are not taken to document it in some fashion is more than metaphorical. Instead, it is the literal fear of disappearance, of oblivion. It is, for example, something particularly prominently manifested in dance, where history shows that the failure to document leads directly to the erasure of dance itself. 'Choreography', writes Ann Hutchinson Guest, 'has been called "the throwaway art" because so many ballets were allowed to be forgotten' (1984: xi). As Fernau Hall also observes, 'To anyone approaching the study of ballet with knowledge of other arts, what stands out most clearly is the poverty of its traditions' (1983: 390). Fragments, scenes and rumours of ballets exist – along with titles, names of choreographers and dancers and the music – but Hall observes how we are able to list only a small number of complete ballets that survive from before the twentieth century. This disappearance of dance has been the result of a lack of comprehensible and enduring medium of documentation or of any easily applicable and accessible form of dance notation (see also Salter, 1978).

Indeed, many dance scholars enviously perceive notation as having provided music with a method of preserving and safeguarding musical history. Notation enables music to be written down, making possible its communication across generations. As musicologist and historian D.W. Krummel puts it: 'Music on paper lasts; it can aspire to immortality, enduring for centuries, conquering the ravages of time. Such has been an article of faith, at least for two centuries and, deteriorating paper notwithstanding, even now' (1987: 6). The long heritage of western music demonstrates the 'literacy' that written notation provides and its success in fulfilling its documentary purpose. In contrast, the difficulty of notating dance has made it 'illiterate'.[7] Even as recently as 1990, a study in the United States declared that there was a 'crisis in the documentation and preservation of dance', suggesting that the lack of widely accepted methods of documentation continued to result in dance leaving few records behind (Dance Heritage Coalition, 2001). The dances not documented disappeared: they needed saving from oblivion.

This very real fear of disappearance, of oblivion, is a dramatic reversal of the celebration of transience constructed and manifested within discourses of disappearance. Again, as with ideas of transience, the articulation of the need for documentation is most prominent amongst those with strong personal and social investments in live performance. As critic Matt Wolff comments 'Theatre people understandably resent that great performances are forgotten by future generations' (in Lawson, 2004). Here transience is no longer a position of strength, of unique power or of political and ideological resistance, but instead becomes something to be feared and combated. As witnessed in the comments of one anonymous Edinburgh Festival Fringe theatre director:

> In five weeks what will be left of [my play]? A script, a press release, a couple of photos, and the reviews. (In Shrum Jnr, 1996: 11)

What is worrying this director is the imminent disappearance of his or her production, the passing of an ephemeral event and the fear that any record will be mere residue and inadequate remembrance. The result is that positive value and the very ability to say something of performance is dependent upon the act of retention. If disappearance evokes a radical ontology then it also seems to speak of oblivion, forgetfulness and death.

Documenting the ephemeral

Those wanting to save performance, therefore, deem documentation an end in its own right, a quasi-moral duty, with any subsequent equivocations or motivations largely addenda. Fear of transience, of the forgotten equating to the valueless, has long sparked a practical, social and academic urge to 'save' live performance from disappearance. In direct contrast, Jane Blocker is inspired by Phelan's manifesto of disappearance to declare that performance is the antithesis of 'saving' (in Schneider 2001: 100). Strikingly, articulations of both the value of transience and the duty of documentation possess their own political and moral imperatives and perceptions.

This shared moral language also shares a motivating impulse, as documentation and disappearance are both fundamentally driven by the positive valuation of performance as live. For example, McAuley echoes sentiments about the need to save performances and also details the need to persuade sometimes sceptical practitioners to take responsibility for their 'legacy': 'While some individuals may feel anguish at the lack of more durable traces of these experiences, most theatre artists are more interested in their next show than putting resources into documenting the one that has just closed (1994: 184). At the same time, however, with echoes of Phelan she also articulates the value of performance in terms of the discourse of transience. 'Theatre, by its nature,' writes McAuley:

> is an art of the present moment, and the theatre artists focus their energies on the present of the lived experience. Performance is unrepeatable and is fascinating to performers and audiences precisely because it is unique and ephemeral. (1994: 184)

The shift that McAuley operates here – from the positive valuation of disappearance as central to performance, to the subject of documentation – is a movement that quickly becomes familiar when reading around discourses of disappearance. It is present, for instance, in the contradiction that dance critic Marcia Siegel notes when she describes 'the critic's paradoxical passion to want to capture the moment and simultaneously let it go' (1991: xvi). Similarly, Michael Kirby first defines performance by transience and then declares the importance of halting disappearance: 'The need for performance documentation lies in the nature of theatre itself. Unlike the other arts, performance is perishable' (1974b: i). At the same time as transience is valued as a positive marker of the power of performance, this ephemeralilty is also bemoaned and regretted. In a simultaneous valuation, disappearance and documentation go hand in hand.

These tensions and contradictions permeate this entire issue. Existence *only* in the here and now, ephemerality, is the very thing valued about performance. Even those fiercely fighting for documentation continue to define performance in terms of transience. However, since documentation compromises the existence of performance *only* in the here and now, such documentation and the positive valuation of it erode the definition of liveness. Patrice Pavis neatly elucidates this problem, suggesting that 'we always have an uneasy conscience when notating the theatre, as though we were carrying out a forbidden act which makes the very object supposed to be re-presented, disappear' (1982: 129). Equally, it is telling that Phelan writes of the reproduction of performance as a 'betrayal', while Marco de Marinis' and Annabelle Melzer's explorations of the video documentation of performance are titled respectively 'A Faithful Betrayal' (de Marinis, 1985) and 'Best Betrayal' (Melzer, 1995a, 1995b). For if what is essentially valuable about performance is its liveness, manifested in disappearance, then this would appear to be negated (or betrayed) by documentation.

To expand on this point, it is worth returning to Walter Benjamin and particularly his concept of 'aura', which he describes as the quality perceived in a work of art established by its distance and uniqueness.

Benjamin contrasts the limited and exclusive audience of a non-mechanically reproduced work of art with the potentially limitless audience of the mechanically reproduced work (1970: 211–44). Erosion of distance, Benjamin declares, erodes the 'aura' of art. In discourses of performance, the idea of 'disappearance' directly parallels that of 'distance', with the valuation 'live' (standing in for 'aura') seen as a function of 'disappearance'. Therefore, any eroding of disappearance erodes the 'live'. The relationship is clear in Genet's statement of the limited audience of a theatre production: 'It will not be possible for all the living, the dead, and the future generations to see *Les Paravents*'. These ideas are repeated again, once more echoing Benjamin, in Phelan's declaration that 'Performance honors the idea that a limited number of people in a specific time/space frame can have an experience of value which leaves no visible trace afterwards' (1993: 149). However, just as Benjamin argues that mechanical reproduction has enhanced perceptions of aura in manually produced works, so it would be a mistake to believe that documentation negates disappearance. Instead, it is possible that some degree of retention heightens perceptions of ephemerality; the two are certainly not exclusive.

Technology and disappearance

Much of the discussion within discourses of disappearance seeks to describe *what* performance is. Indeed, Phelan is engaged in an explicit attempt to describe the ontology of performance as 'representation without reproduction'. The difficulty, as demonstrated in the tension existing between documentation and disappearance, is that without a stable conclusion the entire argument is prone to crumble – by their nature, ontological arguments do not cope well with internal contradictions and the maintaining of paradox. In the ongoing conceptual debate over the nature of liveness, it is precisely such insertion of doubt and contradiction that Auslander seeks when he questions the notion that any live performance can exist without entering into documentation, reproduction, repetition and mediatization. Ontological formulations of performance may contrast the disappearance of live performance with the automatic reproducibility of technological performance, but Auslander suggests our cultural impulse to replicate and repeat has rendered any such distinction irrelevant (1999: 50).[8]

If technology has a central position in defining and delineating ideas of disappearance and liveness, then the existence and abilities of recording technology – photography, CD, video, DVD and so on – are also at the core of the urge to document. McAuley suggests that the comparison of live arts with technological media has been the primary motivating factor in the demand for and interest in performance documentation (1986: 5). In other words, and echoing Auslander, she suggests that there was no overriding desire to document performance before mechanical forms of recording became familiar, not least because such documentation would have been impossible. Theatre historian Laurence Senelick similarly considers whether the desire to record performances was a result of technological developments, or if technology instead enabled the fulfilment of a pre-existing need. His suggestion is that such desire is probably the 'product of mid-nineteenth century positivism' (1997: 256). Indeed, it seems fairly certain that the existence of increasingly sophisticated methods of recording intensify both the urge and the expectation for documentation. That they instigate the very desire is more doubtful, although even Jonson's desire to redeem his plays from oblivion through publication can be seen as dependent on the existence of mechanical printing technologies.

What is clear, however, is that recording technologies do ask fundamentally different questions about the relationship between document (or record) and original than are presented by pre-mechanical possibilities. Indeed, such is the complexity of the relationship between live and technological performances that it is possible to question the notion that any live performance can exist today without documentation, reproduction and repetition. As Baudrillard suggests, our cultural impulse to replicate and repeat has gone a long way to rendering any distinction between original and copy irrelevant. It is against such circumstances, and what she terms 'the "preserve everything", "purchase everything" mentality so central to the art world and to late capitalism more broadly', that Phelan explicitly identified performance's unique power (2003: 295). However, it is also only in terms of such impulses that disappearance becomes meaningful.

Taking such arguments further, it is possible to see performance as defined not by its disappearance but by its inscribed and potentially enduring documentations. The perception that performance disappears is dependent on retention and documentation. Just as Auslander and other commentators declare that there can be no concept of the live without the mediatized, so there can be no concept of ephemerality without documentation, no sense of loss without memory. We can only lament the erasure of dance history and performances past because some rumours of their splendour survive. Consequently, within discourses of disappearance it is necessary to recognise that some continued existence and retention within memory is always at least implied.

However, the logic here must work both ways: there can be no concept of documentation without a sense of that which is not (or cannot be) documented. A documentation that tells the *whole* story is not documentation, but the whole story; not a recording, but the thing itself. Most contemporary 'recorded' music, for example, is not a recording (are not documents) in the sense that it is not a copy of a performance that either did ever happen or indeed often could ever happen. They are not recordings or documentations but instead are the thing itself and the whole thing, with nothing missing. By contrast, a documentation of a live event is partial and incomplete. Consequently, that which is missing (the unrepresented, unrepresentable and liminal) re-inscribes the continuing absence of the ephemeral performance. The discourse of documentation continually re-inscribes perceptions of ephemerality; the act of documentation marks and brings into being the fact of disappearance.

Disappearance and documentation

The documentation of practice forms an ongoing preoccupation within performance studies. Faced with the need to see, know and think about performance beyond its transient existence in performance, practitioners and theorists alike have grappled with the dilemma of documentation. As such, documentation is both the product of disappearance and in turn also productive of disappearance. The apparently contradictory discourses of disappearance and documentation are inherently interdependent, self-fulfilling and self-perpetuating. Some degree of essential retention prompts the very idea of disappearance; while the partiality and incompleteness of any documentation returns us to the transience of the performance itself.

Emerging from this continual pairing and repeated double bind, the following chapters explore what might be considered the space created within and between documentation and disappearance, looking at how various activities and media of performance representation respond to and reflect upon the competing practical and ideological demands presented by these discourses. The chapters explore how documentation does not halt disappearance, but contains within it a memory of transience. They explore how what is shown and said also returns to reinstitute what is left unsaid and unshown. At the same time, however, disappearance would be mute, unknown and unknowable, without some form of trace or residue. The representation of performance forms a kind of audiencing of performance, making it seen and giving it voice and memory. In this manner, the act of documentation makes disappearance visible.

Part I

2
Archives

Within discourses of documentation, nowhere can the impulse to stop things from vanishing, and the feeling that one is able to access the past, be stronger than with the archive. Archives are by conception and practice intended to preserve traces of the past, making it available for future generations to access, study and, more broadly, simply to know. To archive is synonym with to document; to archive is to do documentation. To archive symbolically asserts ideas of recording, preserving and remembering events and the past. This is not least the case, although not unproblematically so, with the performing arts archive. It is unsurprising therefore that the question of archives, and theories and hopes about archives, permeate through and around discourses of live performance practice, research and theory. In response to perceptions of transience and disappearance, the archive *in concept* represents the purest desire to document and preserve live performance.

This chapter explores the concept of the archive in and of itself, examining how the manifestos of archival institutions and statements of archival promise form a discourse that constitute a perception and ideology of the archive. Within this discourse, the archive is often constructed as the most 'proper' storehouse of performance afterlife: claiming characteristics and values of authority and legitimacy, frequently asserted according to its unique and national or even international status. Equally problematically, discourses of archival authority and worth also seek to assert properties of completeness, neutrality and objectivity. This chapter explores postmodern challenges to such constructions of archival authority, particularly in terms of the metaphorical relationships constructed between archives and ideas of individual and national memory. At the same time, however, the chapter examines the enduring allure of the archive, a *need* for archives deeply rooted in our cultural practice and identity.

The promise of archives

That archives, or any historical research based upon primary evidence and documents, offer unproblematic, objective or complete access to the past is something that has undergone rigorous theoretical deconstruction and criticism. In particular it has been challenged by concepts of postmodernity, which have systematically disputed any claims to authority or singularity.[1]

The postmodern challenge to archive practice, and to history more generally, has been to examine the constructive role that narrativisation, language and the historical document perform in creating our understanding of the past.[2] Such re-envisioning of the archive has emphasised its dependence on scraps and fragments, highlighted its inherent gaps and stressed the re-creative and re-interpretative nature of the archival endeavour. Consequently, the powerful imagery that declares that the archive reveals the past to us is complemented by equally forceful counterclaims of archival limitation that question any assertion of truth or objectivity. Theorists now stress that the promise to neutral access is based on compromised positions of selection, omission and manipulation. Carolyn Steedman, for example, describes this constructed nature of the archive:

> The Archive is made from the selected and consciously chosen documentation from the past *and* from the mad fragmentations that no one intended to preserve and that just ended up there. ... In the Archive, you can not be shocked at its exclusions, its emptinesses, at what *is not* catalogued. (1998: 67)

Far from being complete – or authentic or objective – Steedman persuasively argues that the archive is the reverse. At the same time as it is possible to point to the sheer wealth and bulk of material in any archive, it is also necessary to acknowledge the even larger body of material not present. Indeed, it is also inherently impossible to say exactly what is missing and where the gaps might be, with such archival fallibility and emptiness inevitable.

Given the prominence and importance of these challenges to ideas of historical truth and stability it is, therefore, wise not to present ideas of archival completeness and authority as a rhetorical straw figure to be emblematically and too easily overturned. This said, a number of writers have very plausibly suggested that there exists a gap between contemporary theory and continuing archival and historical practice. Richard Evans' introduction to his own rebuttal of postmodern attacks on ideas of history, for example, cites Nancy Partner's assertion that the 'theoretical destabilizing of history ... has had no practical effect on academic practice' (in Evans, 1997: 5). And so, while not forgetting the theoretical challenges that have been made to the archive, there does seem to remain a latent, sometimes unconscious but nonetheless powerful, cultural faith in the value of archives and archival research. This can be seen as the culturally produced and perceived promise of archives, which often outstrips both postmodern theory and archival practice.

The nature of this archival promise, as distinct from an archival theory or practice, is presented by Irving Velody who observes that 'As the backdrop to all scholarly research stands the archive. Appeals to ultimate truth, adequacy and plausibility in the work of the humanities and social sciences rest on archival presuppositions' (1998: 1). Our archival hopes and expectations, therefore, are constituted in values of truth and in the possibility of being able to achieve an adequate (or as complete as possible) and plausible (as accurate as possible) reconstruction of the past and past events. As Steedman writes, our attitudes to archives are to do with wanting traces and fragmentary evidence of the past to amount to a comprehensive whole, of wanting:

> things that we put together, collected, collated, named in lists and indices; a place where a whole world, a social order, may be imagined by the recurrence of a name in a register, through a scrap of paper, or some other little piece of flotsam. (1998: 76)

And so indeed, for Richard Evans an archive is a place holding hoards of original manuscripts wherein it is possible, by dint of careful research, to 'reconstruct the past accurately' (1997: 18). Such hopes also hold true for the performing arts archive, as much as for any other, where archival research holds out the promise of reaching back to origins in the attempt to reconstruct and rediscover past performances.

The performing arts archive

This articulation of archival promise is manifested in the mission statements of numerous live performance archives and institutions, each time demonstrating a basic and deeply embedded investment in archival activity as not only an act of documentation but also a form of resistance to disappearance. For example, the Theatre Museum, part of the Victoria and Albert Museum in London, introduces its collections as follows:

> The Theatre Museum exists to increase the enjoyment, understanding and study of the history, craft and practice of the performing arts in Britain through its collections, which are the largest of their type in the world. The Theatre Museum collects a wide range of documents, artefacts and works of art which record the history of the performing arts in Britain from the sixteenth century to the present. (2005)

Also describing itself as providing a 'national record of live performance' and as 'systematically' collecting material, the Theatre Museum locates the value of its activity within a discourse of archival promise and documentary retention. Elsewhere the Billy Rose Theatre Collection, part of the New York Public Library, describes itself in very similar terms as:

> One of the largest and most comprehensive archives devoted to the theatrical arts. Through conservation and documentation efforts, it preserves and promotes the theatre, playing a dynamic role in the national and international theatrical community. ... the Collection's strength and uniqueness lies in its unparalleled collection of theatre ephemera as well as its pioneering efforts to document theatre on videotape and film. Approximately 5 million items illuminate the art of theatre worldwide. (2005)

Again it is possible to unpick this declaration in terms of the archival promise it articulates, particularly looking at the repeated use of words such as 'unique', 'unparalleled' and 'comprehensive'. Such analysis may be inviting but is perhaps too easy: after all such statements are self-descriptions and public manifestos rather than nuanced articulations of archival theory. However, it is also possible to see statements such as these, along with the activity and intent of such archival institutions, as constructing a discourse that re-enacts and re-enforces the powerful cultural promise of archival retention. Indeed, it is a strikingly prevalent discourse, frequently manifested through different organisations: Arts Archives, for instance, is produced by the Arts Documentation Unit in Exeter and 'dedicated to documenting the processes at work within contemporary performing arts practice' (2005), the Live Arts Archive at Nottingham Trent University 'continues to document current events as they occur and seeks to make its historical record as complete as possible' (2000), while the Jerome Robbins Dance Division, a counterpart to the theatre collection at the New York Public Library, describes itself as 'the largest and most comprehensive archive in the world devoted to the documentation of dance' (2005).

In these mission statements the motivation for the archival endeavour rests in the duality of the relationship between documentation (and a widespread cultural impulse to preserve and save) and disappearance (and the perception of the transience of live performance). Indeed, as the Jerome Robbins Dance Division also observes, those working in dance have a particular fear of the disappearance of live performance that is directly linked to the perception that it is a form traditionally very difficult to document. Or, as Joanne Pearson puts it, 'the archival problem of all dances [is] their tendency towards invisibility' (2002).[3] Tellingly, many archival institutions describe themselves as proactive recorders of the performing arts rather than passive repositories – the Theatre Museum, for instance, 'actively documents live performance'; the Jerome Robbins Dance Division 'works to ensure the art form's continuity through an active documentation program' – going out and finding or making documents of performances, rather than collecting them after the fact. Again the implication is clear, the ephemerality of live performance means that it must be consciously documented if it is not to disappear, with the primary preoccupation not the creation of new art but ensuring the documentation of existing art. These are organisations looking to the past and the future rather than the present.

That in seeking to preserve these institutions cast their eyes to the future is another characteristic of archival activity, which is predicated on a consciousness of future audiences and future attention. Indeed, Jacques Derrida suggests that archives allow an idea of the future to emerge, writing that 'the question of the archive is not, we repeat, a question of the past. ... It is a question of the future, the question of the future itself, the question of a response, of a promise and of a responsibility for tomorrow' (1995: 36). The construction of an idea of our present (what the present means, in a sense) is dependent on a process of historicisation that envisages a responsibility for what the future knows of today. In this context, for the archival institutions, performance is constituted with value precisely by being perceived as of value to future generations. This value is only enhanced by the historical fragility, the transience of performance and the deliberate effort that has to be invested in its preservation.

Read collectively, the mission statements of these archival institutions construct a discourse that valorises archival usefulness and purpose, and in doing so reiterates the promise of the archive anew. Present within this discourse is an emphasis on qualities that re-state the nature of archival worth, including assertions of the size and uniqueness of the collection and statements of its national or international importance.

The message of these archival institutions is evident: we should place documentation at the centre of creation itself so that as work is performed it is recorded. As a discourse constructed in a shared vocabulary, these mission statements articulate clear perceptions about the value of archival practice along with the transformation of the positive valuation of the ephemerality of live performance into a fear of ephemerality and a subsequent valuation of documentation, the document and the archive. (Yet at the same time, of course, the inherent value of the document is dependent on qualities of disappearance.) There is a quasi-moral dimension to this ambition, evident in the language emerging in the discussion: performance must be 'saved' or 'rescued'; it is part of our 'heritage', our 'legacy' and must not be 'lost'. As a moral endeavour, the documentation of performance needs no justification beyond these very aspirations. The value of the archive is in the action of archiving, in the act of halting disappearance and preserving for the future; the promise of the archives is that, in halting disappearance, the past – here the performances of the past – will be accessible to the future.

Things, but not the thing itself

The discourse of performing arts archival institutions presents the value of the archive as located in its status as a repository of past performances. This is a promise constructed in terms of a moral objective to retain and protect our heritage and history for the future. What this means in practice, in terms of what is retained, what is documented and what such archival traces say of performance is inevitably far less clear-cut.

As storehouses of the past, museums can be primarily understood as repositories of historical objects – vases, weapons, clothes, statues and so on. Similarly archives are collections of actual, original documents – letters, books, manuscripts, files and so on. Indeed, the primary understanding of archives is often as a written and textual collection, reflecting the primacy of the written word in much of western culture. Either way, in each occasion the museum or archive contains (retains, preserves) the 'thing' itself, with their value resting on this possession of the original object or original document. Given the transience of live art, the live performance archive or museum is more problematic, as it by definition cannot contain actual performances – the thing itself is always absent. As Peggy Phelan writes, 'we have created and studied a discipline based upon that which disappears' (Phelan and Lane, 1998: 8), a statement that

Rebecca Schneider takes further, arguing how 'According to the logic of the archive, performance is that which does not remain ... performance appears to challenge object status and seems to refuse the archive its privileged "savable" original' (2001: 100–1).[4]

Instead of containing the original thing itself, therefore, the performing arts archive represents the formal collecting, cataloguing, preserving and consecrating of traces of past performances, but crucially not the performances themselves. Such performance archives can consist of almost anything, including but not limited to theatre programmes, brochures, leaflets, photographs, videos and sound recordings, press releases and press cuttings, details of marketing strategies, figures of tickets sales, contracts with performers and confidential budgets, correspondence, descriptions of sponsorship arrangements, venue plans, set and costume designs, stage and lighting plans, production notes, playscripts (sometimes annotated), interviews with directors or actors, actual costumes and examples of stage properties and so on. Anything that is remotely associated with the performance can belong in an archive, including material detailing the processes of creation, production and reception.

The documentary merit and use of each of these archival traces of performance warrants consideration in its own right – and in the course of this book some, although inevitably not all, will be considered in detail. None of these, however, can be considered to be the 'thing' itself, and instead the performing arts archive makes physical the dichotomy of documentation and disappearance: preserved traces, but not complete presence; fragmentation, but not complete disappearance. Indeed, the process of sifting, checking, triangulating and validating across these different fragmentary sources can be seen as the essential stuff of historiology, as the Theatre Museum describes it 'Costumes, designs, manuscripts, books, video recordings ... posters and paintings all play their part in helping to reconstruct the details of past performances and the lives of performers, past and contemporary' (2005). Such triangulation is also the approach suggested by Gay McAuley, who argues that the incompleteness or partiality of any one medium or document can be buttressed by information gathered from another (1994).

Despite, therefore, the problematic relationship between the performing arts archive and the thing ostensibly preserved but inherently absent, the construction, use and perception of the live art archive remains deeply rooted in the heart of repeated articulations about the value of archival activity and the usefulness of collecting and examining historical documents and objects. This is clear from the manner Mindy

Aloff describes archives in an article entitled, revealingly, 'It's Not Ephemera After All':

> Although it is customary to speak about dancing as an ephemeral art that leaves nothing behind except the memory of its performance, in fact it leaves much more than you might guess: costumes and sets, musical scores, perhaps notation of the choreography, programs and reviews, photographs, letters, films, and, nowadays, hours and hours of videocassette recordings. While such leavings constitute a husk of dancing, they are also the kernels of dance history. (2001)

Here Aloff constitutes the archive as our memory, our heritage and our best access to the live performance of the past – although in a description of 'kernels' and 'husks' she also recognises the incompleteness of such leavings, residues that include many things except the thing itself. As Aloff's comments also illustrate, the idea that the archive preserves 'our theatre history' or 'dance heritage' continues the discourse running through the manifestos of archival institutions and practice. Indeed, the perception that permanent records of live performance are metaphorical replacements for fragile human memory is a prominent and lasting element of discourses of documentation, and one that will be returned to in Chapter 3.

Stones talk!

Present within the discourses of archival promise exists a continuing duality, between assertions of retention and yet also of loss and disappearance – particularly in terms of references to scraps and fragments and within metaphors of discovery and reconstruction. It is this promise of accurate, truthful and *singular* reconstruction of the past that has been subject to greatest theoretical challenge, questioning the use and conceptualisation of the archive in a manner that has returned attention from the act of documentation back to the action of disappearance.

Something akin to these passions and desires present within the discourses of documentation and disappearance are explored by Jacques Derrida in *Archive Fever*, an elusive work that has fuelled much recent discussion of archive theory. In this book Derrida describes the archive, and an archival drive or fever, as located at a point of loss, disappearance and death, 'There would indeed be no archive desire without the radical finitude, without the possibility of a forgetfulness' (1995: 19). Derrida's concern, therefore, is not with the political meaning of the archive, but instead with its emotional and cultural meaning. And with archives located at and produced by forgetfulness, the meaning of archives can in many ways be found in their motivation, in a fever for archives:

> We are *en mal d'archive*: in need of archives. ... It is to burn with a passion. It is never to rest, interminably, from searching for the archive right where it slips away. It is to run after the archive, even if there's too much of it, right where something in it anarchives itself. It is to have a compulsive, repetitive, and nostalgic desire for the archive, an irrepressible desire to return to the origin, a homesickness, a nostalgia for the return to the most archaic place of absolute commencement. (Derrida, 1995: 91)

In many ways this description of passions, hopes and desires matches those produced within discourses of performance documentation and disappearance, where there is similarly evidence of a burning passion to return to origins in documentation that is continually fighting slippage into disappearance. The cultural urge to document performance, emerging from the self-destructive ephemerality of performance, is a fever for archives produced by an ever-deferring chase for a piece of the origin, for something of the beginning. Derrida identifies the cultural promise of the archive as being to halt finitude and therefore as also always containing within it a kind of end, a kind of 'death drive'. Again this is a description that is appropriate for performance, where disappearance is both welcomed and feared and where the very word 'ephemeral' points towards disease, death and to the radical finitude that is transience.

Within this fever for archives, Derrida also describes what can appear to be a process of 'outbidding' in the attempt to return to or retain origins or 'commencement'. This is a process that shifts between concepts of archives, archaeology and live memory – shifting from original documents, to the site of excavation, to the site of experience – with each offering greater primacy in the attempt to return to 'live origin'. With performance such hierarchy of origins places greatest primacy on the voice of the artist and the value of having been *there* live. Cautioning against such claims for greater originary primacy, Derrida describes this as a process that seeks to efface the archivist (and thereby the sign of interpretation, mediation and translation) and instead assert that 'the origin then speaks by itself ... It presents itself and comments on itself by itself. "Stones talk!"'¹ (1995: 95). And of course stones, axiomically, do not speak for themselves but are instead spoken for – not least, in the case of the performing arts archive, by the voices and institutional authority of the academy, the subject of the next chapter. What Derrida's warning should also prompt us towards, however, is the equally problematic trust and hope that the origins of performance (the performance itself being the stones themselves) can present itself and comment on itself, and thereby provide a superior kind of truth and evidence that avoids the mediating effects of representation and the fragmentation of the archive. Being *there*, like the thing itself, does not speak for itself. Instead it is the archive, along with the various representations of performance contained within the archive, which give performance form and meaning and that speak about performance.

3
Proper Research, Improper Memory

Following on from the exploration of archives and archival promise in Chapter 2, this chapter examines the 'academy' as a second and connected forum or institution of documentary drive and activity. In this context, both archives and the academy are interested in the documentation of performance history; both able to claim, in part as a result of their production by experts and their institutional status, a powerful degree of authority and legitimacy. Discourses of archival worth promise access to an authentic memory of past performances; meanwhile the academy claims authority in scholarship and primary, archival research. If archives are the storehouse of performance history, then it is the respect for expert knowledge invested in the academy that can direct what gets looked at, remembered and recorded in the first place. Further, it is often the academy that invests meaning into (and which speaks for) the documentary traces of performances.

This chapter considers the possible impact of academic attention on the writing and rewriting of performance history, the construction of a performance canon and, more immediately, on directing funding and determining the existence of performance genres. While putting such political implications of the academic documentation to the fore, discussion also explores what idea or meaning of live performance is constructed within academic representations.

Research and the impulse to document

As explored in the previous chapter, despite powerful attacks on ideas of archival authenticity and authority, there continues to be a willingness to embrace the cultural ideal and promise of archives to enable us to access the past. As Derrida describes it, there is a need for archives that is more than pragmatic but also emotional. In the context of the performing arts this archival need is manifested in a fever for documentation, existing even at the same time as the repeated assertion of value and power of disappearance.

One reason for what several researchers have described as the continued allure of the archive is a sensual, almost sexual (certainly fetishistic) satisfaction that comes from the activity of archival research. The ability to touch documents and objects from the past is, in itself, a primal attraction of the archive. Harriet Bradley, for example, stresses the 'pleasures, seductions and illusions of archival work' and the 'intoxication of the archive' (1999: 113); while Helen Freshwater draws on her experience working in theatre archives to describe the 'allure of the archive' as in part voyeuristic pleasure and in part a sense of accessing authentic material and conducting original research (2003: 732–5). This seductive promise of originality and authenticity, constructed by archival discourses and projected in the manifestos of archival institutions, continues to enchant even while most contemporary archive theorists – including Bradley and Freshwater themselves – have deconstructed our understanding of archival documents and historical truth.

Even as researchers have interrogated ideas of archival authority they have at the same time also identified (and thereby themselves added to) a discourse that constructs the archival promise as an ideal – albeit one now acknowledged to be unattainable, nostalgic and emotional rather than rational in origin. Additionally, for the academy and the scholarly researcher, the archive in practice, in experience, continues to provide seductive rewards rooted in perceptions of worth, legitimacy and originality. Certainly the academy and archive go hand in hand, one producing and feeding upon the other. In performance studies, as in other disciplines in the humanities, research often rests upon explorations of archival material.[1] In turn academic expertise supports and directs the collecting policies of many archival institutions, something illustrated not least by the presence of numerous performing arts archives within universities and academic departments. Performance studies as a discipline has also long been interested in pushing the development and testing the usefulness of different approaches to documentation, with examples here including the work of the Performance Studies Department at the University of Sydney and the existence of SIBMAS (the International Association of Libraries and Museums of the Performing Arts) with their ambition of 'promoting research, practical and theoretical, in the documentation of the performing arts' (2005). This interest in archival practices has been emphasised more recently, particularly in the United

Kingdom, as a result of the growth of interest in practice-based research approaches and the corresponding attention being paid to the documentation and dissemination of such work in publication, to meet the requirements of research assessment exercises and to provide satisfaction for funding bodies.[2] Given such need to make the evidence of performance focused research available in other times and places, the academy has often focused on the creation, assessment and evaluation of documents of performance. Indeed, it is clear that academic research into contemporary performance has become not only archival in origin, but also often becomes the act of documentation (the act of archiving) in itself.

One fairly typical example of such practice is Geraldine Cousin's book, *Recording Women: A Documentation of Six Theatre Productions* (2000), which sets out to record performance events otherwise 'subject to erasure'. The belief that the work she is documenting, being by women, is particularly subject to neglect and oblivion partly motivates Cousin. As she notes in her introduction, she records because of a fear that the productions would otherwise 'soon be forgotten'. Her documentary ambition has, therefore, a cultural and social impulse not explicitly related to the works' status as live performances. However, the liveness of the performances prescribes erasure into their very creation: whatever their origin or subject matter they disappear unless actively represented. This instinctive determination to document because the works are performance based is strongly evident in Cousin's expression of intent.

> My aim throughout has been to preserve what could be preserved in book form of these six theatrical events – to provide traces, at least, of powerful, moving and, at times, very funny experiences. There is an anomaly in this, of course. Theatre (as I have already noted) is ephemeral. Play texts, reviews, photographs etc., survive, but the performances themselves are over; they had existence only in the present moment of theatre. One of the roles of the theatre academic however is, I think, to bear witness to what has been, and this is what I have tried to do. (2000: 3)

Cousin, therefore, continues and embraces the description of performance as ephemeral: her expression of theatre's disappearance in the moment of its creation echoes the discourses observed elsewhere in this book. The crucial element, however, is that the positive appreciation of presentness is not matched by a positive valuation of disappearance. Instead, as is demonstrated by the value loaded language she employs, for Cousin the documentation of performance is a quasi-moral endeavour, the action being its own reward as a good thing in its own right. The evident tension between the valuation of presentness and the desire for retention through documentation is unexplained. Again these two elements – documentation and disappearance – are tied together in a prominently repeating double bind.

Other questions also leap out unanswered from Cousin's statement of intent, particularly the attention she pays to her role as an 'academic' and the note that she is preserving what can be retained 'in book form'. Partly apologetic for the inability of writing to account for the performative nature of theatre, this also presents an unconscious acceptance of books and scholarship as the right and proper place for such things to exist. Cousin's reference to 'book form' is echoed more knowingly elsewhere, when Clifford McLuhan of Brith Gof introduces his own documentation of the company's *Tri Bywyd* with the observation that 'the limitations of page by page turning, of monochrome print and of size [are] all necessary to link format and academic worth' (in Kaye, 2000: 125). In other words, existence in print, in book form, is paramount for existence in the world of the academy. Discussing this attitude to the 'existence' of performance history only in publication, Anna Cutler describes the privileging of what she calls 'Proper Documentation', 'defined here as the material related to a performance which is signified by the written word and made available through publication' (1998: 112).

Such documentation, Cutler declares, attempts to claim the status that 'if it's not printed, it doesn't exist', which she sees as resulting in the neglect of work outside of scholarly exchange. And, indeed, Cousin's work is part of a prominent and widely echoed scholarly discourse that articulates the desire to produce and publish documentations of live performance, thereby introducing it into critical exchange and providing performance not only with a permanence but also the passport of cultural literacy that comes with paper and the written word. In this discourse of documentation, expressions of self-validation and the value of existence in publication soon become familiar, as, for example, in Greg Giesekam's account of Clanjamfrie's *The World's Edge*:

> While it may be like looking at a snake's old skin after it has sloughed it off and moved on, we should recognise that the dearth of descriptive accounts contributes to the erasure of such approaches to performance from most published treatment of recent British theatre. (1994: 115)

As with Cousin, it is worth noting that Giesekam is partly motivated by the perception that the work under consideration here, non-text based performance art, is particularly subject to erasure. However, it is the perception and fear of the disappearance of live performance that is a more fundamental motivation. Giesekam, like Cousin, also acknowledges the disappointments with paper documentations but operates the same valuation of the published over the unwritten and the studied over the unstudied. Once more, all that is stated is that we (by implication the academy) must document because, if we do not, a performance cannot continue to exist. Both Cousin's and Giesekam's assumption is that to exist in publication is to escape erasure. Further, existence in scholarly exchange is implicitly defined as a positive status that establishes the value of its subject as a direct result of its existence. This assumption is made explicit by archivist Michelle Potter, for whom the desire to document dance is grounded in the fear that 'without efforts to preserve the history and heritage of the art form it will forever languish as trivial and not worthy of serious research' (2000). Echoing these sentiments is Keir Elam's observation that the stage spectacle has long been considered 'too ephemeral a phenomenon for systematic study' (1980: 5). Even more unequivocally, Denise Varney and Rachel Fensham declare that live performance must be documented to ensure that it 'is included in contemporary critical discourse', as otherwise it 'will become increasingly absent from critical theory' (2000: 96).

Proper research

Such statements construct a correspondence and mutual dependency between documentation, scholarship and legitimacy. They present a self-reflective fear that the ephemeral must be documented, lest it be considered ephemera. And alongside this a circularity in terms of what gets documented and gets our attention: if it is important, we must document it; if it is documented, it must be important. The value placed on study and scholarship here, on expert knowledge and the academy, can also be seen as the implicit position and unspoken belief resident within all discourses of documentation and as such is worth exploring further.

The discourses surrounding the academic documentation of live performance demonstrate a concern amongst commentators that if something cannot be touched or measured, examined or judged then it (somewhat paradoxically) both does not exist and worries us. That the undocumented does not exist is evident in the scholarly self-valuation of the studied, with the desperation of performance scholars to draw their subject matter into the 'researchable' demonstrating the relation of the unscholarly with the unvalued. The undocumented history of performance is unknowable, a vacuum in our knowledge that motivates a desire to save performance from its self-destruction in disappearance. That this is motivated by a fear of disappearance is more speculative but demonstrated in the moral dimension that is present within the language of this discourse.

Also demonstrating the fear of disappearance is the indignant condemnation by some writers of any positive value placed on 'liveness' as un-documentable and therefore unknowable. Elements that are valued as live, including presence but particularly here disappearance, are things that cannot be measured empirically and as such warrant automatic suspicion. As Caroline Rye writes

> In our present-day virtually wired-up digital world we have become doubtful of the value of time and space specific, oral, aural, unmediatized, multi-sensory and in some senses un-recordable encounters which characterize the interactions which occurs when groups of people meet together to partake in performance. (2003)

Evidence of the suspicion of the non-proper or non-empirical knowledge of live performance includes the strident rejection by Auslander of all ontological understandings of live performance as being merely mystical, magical clichés (1999: 2). Elsewhere, Roger Copeland terms the valuation of performance's liveness 'sheer bourgeois sentimentality' (1990: 42), while Varney and Fensham write of the 'reactionary metaphysics of presence' (2000: 96). Documentation firmly places live performance in the realms of the known, the empirical, in the realms of 'serious' consideration; this again is a moral struggle against the irrational or essentialistic and against any belief in the mystical or ineffable.[3]

In interrogating the allure of the archive to the academic researcher, Helen Freshwater writes that 'It seems that the temptation of making a claim to the academic authority conferred by undertaking "proper research" may prove irresistible for the researcher utilizing archival material' (2003: 731). In other words, by documenting live performance we enable proof, authentication, evidence and study. By documenting performance, we halt – or seek to halt – disappearance and also doubt, instability, multiplicity and ineffability. Documentation as proper research establishes – or seeks to establish – a firm grasp upon the performative.

The wider implications of this valuation of scholarship, and of the very existence of performance studies as a discipline within the academy, are difficult to quantify. However, as with the power plays implicit within the archive drive, so does the relationship between the academy and performance practice have both a potential and actual impact not only on the construction of performance history but also on its unfolding present. Academic researchers are aware of this when they assert the value of their work in giving voice and attention to performances that otherwise might be forgotten or neglected. The attention on performances from marginalised communities, whether by gender, class, race or politics, is precisely motivated by the desire to give them voice and the prominence and prestige than comes with academic reflection. In other words, the work of academic documentation and archival construction is also the work of constructing and shaping a performance 'canon'. The positioning of works within a recognised canon marks them as important, and not insignificantly as saved from disappearance. Yet if to be in the canon is to exist and endure, then this also, by definition, allows only a narrow body of work to be saved.

Observing the implications of this activity in practice, Jill Davis writes of how theatre historians and theorists has successfully sought over the years to construct 'feminist theatre as an academically "proper" discipline' – echoing the strikingly evaluative term 'proper' already used in this chapter by Cutler and Freshwater. This is a process that has resulted in feminist theatre being defined by academic writing and naming, and the institution of a relatively narrow canon of performances deemed worthy of serious attention. Indeed, Davis suggests that the current canon of feminist theatre studies is 'small, mostly American, and focuses on performance art rather than other forms of theatre making'. In practice this results, Davis suggests, in the

> Repeated return to the same small set of theatrical objects (and it might be noted that it is not only the same performers, but the same performances, some given a long time ago). (1999: 185)

This situation is perpetuated in the same way in which the canon perpetuates itself: with critical study or documentation begetting further critical study. The performances themselves become commodities in academic exchange.[4]

Another example of this kind of impact of performance documentation on performance practice and history is presented by Helen Iball, who moves away from high-level academic research to note similar connections within student teaching and learning. As Iball notes, even with contemporary performance courses, students are unlikely to have seen much of the material studied in actual live performance, the result being an added emphasis on the existence of work in and through documentation.

Using the website for British physical theatre company DV8 as her main example – www.dv8.co.uk, which includes a prominent link titled 'looking for info for your project/dissertation' that leads to an impressively comprehensive 'virtual information pack' and archive – Iball describes how increasingly performance companies and practitioners 'are expected (by the student) to take some responsibility for coursework generated by the performance company's absorption into the educational canon' (2002: 60). In other words, and rather deliciously, from the practitioner's perspective both the reward *of* and requirement *for* canonical status is documentation. Without documentation the work is unknowable to study and therefore absent from the critical canon.

The narratives presented by Davis and Iball describe the power relationships between performance scholarship and performance practice, replicating in many ways the questions about political power, memory and democracy raised through archive theory. It is evident that academic attention inevitably privileges writing and paper-based reflection over other possible responses to and reflections on performance. It is also apparent that academic attention begets further academic attention. More speculatively, just as scholarly attention has an impact on what gets studied and what enters the historical archive, so too does academic attention have the potential to impact on what performances and companies get the respect, consideration and media notice that allows them to successfully attract audiences and obtain funding through state and other subsidies (this possibility is discussed further in relation to the work of Forced Entertainment in Chapter 4).

In the performing arts, discourses of archival authority present the academy with the opportunity for 'proper research': proper in being both authentic and authorised and in claims to validity beyond the anecdotal or speculative. In other words it is very tempting to believe that having done the archival legwork researchers are able to assert the legitimacy of their arguments, thereby trumping more speculative or theoretically-based arguments. This is the attraction of the archive for the performing arts researcher, where, as each performance disappears, it offers the possibility of supplementing and perhaps supplanting doubtful (and inherently liminal) memory as the site of performance record. As theatre historian Robert Erenstein notes, the theatre researcher 'needs documents to justify his field of research [and] must resign himself to the ephemeral nature of the performance; once ended, it lives on only in documents and in the memory' (1997: 185).

Improper memory

This comparison of the archive with human memory is a popular motif in contemporary archive theory. Carolyn Steedman, for example, notes a common desire to use the archive as a metaphor for memory. This observation is again present when Richard Harvey Brown and Beth Davis-Brown explore the role the archive plays in defining national memory and consciousness (Brown and Davis-Brown, 1998: 17–32). Irving Velody also examines the idea that the 'modern memory is now above all archival' (1998: 13).

The metaphorical or literal alignment of archive with memory is particularly relevant to the performing arts, especially when examined alongside radical declarations that the only trace of the live event can and should be the audience's memory. Peter Brook powerfully suggests this, declaring that the only witnesses to a performance event are the people present, 'the only record is what they retained, which is how it should be in theatre' (in Melzer, 1995a: 148); while Patrice Pavis writes that 'The only memory which one can preserve [of live performance] is that of the spectator's more or less distracted perception' (1992: 67). The cultural value and attributes placed upon such originary memory can be significant. As Derrida discusses in *Archive Fever*, the archival promise of live memory seems to outbid that of the archive itself, promising the primacy of being an origin that speaks for itself. However, comparison of conceptions of audience memory with those of a theoretical archival memory soon reveals that they articulate very different perceptions of the objectives and value of any retention of ephemeral performance.

The enabling of a more accurate and accessible memory of live performance is one possibility articulated by discourses of archival authority. Indeed, the perception that permanent records of live performance are metaphorical replacements for fragile human memory is a prominent and lasting element of discourses of documentation. For example, in 1913 a review in *T.P.'s Weekly* of a book on the Russian Ballet declared:

> The one glaring fault of the Russian Ballet is that it has passed away with the brief traffic of the stage, as is the case with all theatre work. Now, however, comes a chance of chaining the vision, of retaining links of memory to bind us to the dreams of dance, to keep forever in our view ... The possessor of this book has the Ballet in epitome. (Anonymous, 1913: 764)

Many of the ideas articulated in discourses of documentation and disappearance are echoed here: including the perception of performance as ephemeral, the desire to halt transience and use of memory as a metaphor for archival retention. Today, comparisons with mechanical methods of reproduction make the romanticised illustrations included in the book *The Russian Ballet* seem wholly inadequate and inauthentic as documentation of the performance. However, while expectations and standards have changed as the result of technological developments, the motivation to retain transient performance in some form is the same. The result is that, almost a hundred years later, this discourse of documentation and retention remains startlingly similar in terms of its language and articulation of values. So by way of contemporary comparison, in an article in *The Village Voice*, theatre reviewer Michael Feingold praises the services of the commercial Broadway Theatre Archive as offering 'videotaped memories' of theatrical performances of the past (2000). Similarly, Marcia Siegel sees a major responsibility of the critic as being the 'memory' of performance (a theme developed further in Chapters 9 and 10), writing:

> By that I don't mean that critics have the best memories ... I mean that they are professional observers, and that what they tell us is our only systematic account of an ongoing history. (1977: xiv)

This relationship between live performance, documentation or archival collection and memory also comes to the fore in Varney and Fensham's discussion of the importance of video recordings of theatre performances to academic knowledge and research (the particular nature of the video record of performance is the subject of Chapters 5 and 6). Here the ambition is essentially the archival objective of 'saving' live performance, which is formulated in active opposition to any positive valuation of audience memory. To this end Varney and Fensham declare: 'Surely the very ephemerality of individual memories should make it suspect as a reliable record for a performance truth' (2000: 91). It is possible to construct such proper documentations, therefore, as our *proper memory* of performance, superior to actual but improper memory in terms of its accessibility, durability and objectivity.

Taken to extremes, arguments asserting the fundamental need (prerequisite even) for documentation seem to suggest that our performing arts history is what exists in documents and *only* what exists in documents. As such the documentation, and the academic reflection that it allows, does not aid memory but replaces it. The original experience, which for

Brook and Pavis exists in the audience's memory, becomes devalued in comparison as subjective, inaccessible and disappearing. However, the perception of memory as a valuable trace of live performance is more than simply a description – it is not perceived as a problem best overcome by employment of a better 'memory' such as the archive – but is also a positive valuation. Comments by Eugenio Barba illustrate this:

> The spectator does not consume these performances. Often s/he does not understand them or does not know how to evaluate them. But s/he continues to have a dialog with the memories which these performances have sown deep in his/her spirit. I say this not as a director but on the basis of my experience as a spectator. (1990: 97)

This is what Barba means when he writes elsewhere that 'theatrical performance resists time not by being frozen in a recording but by transforming itself' and that such transformations are found in the memories of individual spectators. Barba does not value audience memory despite the transformations it enacts but because of them. Memory, he argues, is in this transformative, multiple and mobile nature closer to the essential identity of the live performance after, not before, it has undergone such transformations. Consequently, if we value performance in terms of its time-based transience, its disappearance, then memory must be a more appropriate site for any trace or afterlife than the frozen and unchanging archive.[5] The archive or video recording may claim to show things as they really were, but Barba declares that the performance is not really what *was* happening on stage but what *is* happening in the minds and subsequently the memories of the audience:

> In the age of electronic memory, of films, and of reproducibility, theatre performance also defines itself through the work that living memory, which is not museum but metamorphosis, is obliged to do. (1992: 78)

Those who object to the positive valuation of the memory as a legitimate trace of live performance do so because of the subjective, inaccessible and transformative nature of memory. Marvin Carlson, for example, writes that:

> Even those fortunate enough to witness the original are unable to return to it to check the accuracy of their memory or to test subsequent hypotheses against it, and for others there remains only

the thinner substance of an experience filtered through the selective consciousness and reportage of intermediaries. (In Melzer, 1995a)

Varney and Fensham echo this assertion of the value of proper research when they express vehement distrust of any positive valuation of subjective memory, condemning it on several counts, including elitism, unacknowledged selection and lack of detail or accountability. Memory, they state, does not 'produce a purer form of truth' (2000: 92). This, however, depends on what kind of truth about live performance one is attempting to reach, and what it is about live performance that one is attempting to 'produce' (or perhaps reproduce).

Most intriguingly, the metaphorical relationship constructed between archive and memory is *more* appropriate as a result of contemporary understandings of archives as unstable, as 'read into' rather than read, than for any hypothetical ideal of the authoritative archive. If each remembrance re-creates memory, if memory is inherently transformative, then so is the archive's construction of the past re-created each time it is accessed.

> An Archive is not very much like memory, and is not at all like the unconscious mind. An Archive may indeed take in stuff, heterogeneous, undifferentiated stuff ... texts, documents, data ... And order them by the principles of unification and classification. This stuff, reordered, remade, then emerges – some would say like memory – when someone needs to find it, or just simply needs it, for new and current purposes. But in actual Archives, though the bundles may be mountainous, there isn't in fact very much there. The Archive is not potentially made up of everything, as is human memory; and it is not the fathomless and timeless place in which nothing goes away, as is the unconscious. (Steedman, 1988: 66)

The possibility that there is more, rather than less, in memory is intriguing, pointing towards a hypothetical archive of the unconscious, located within audience memories, in their conversations and in the emotional reactions and experiential repercussions of the performance. Such an archive would take contemporary archive theory, and the positive valuation of memory, and embrace the transformative conditions of both memory and archive. This collective memory of performance might be located in oral history and investigated through formal audience research exercises[6] or alternatively located in found sources, such as diaries, letters, commonplace books, theatre clubs or internet discussion boards. To perceive such problematic, sometimes nebulous and contradictory sources as containing *more* would be to challenge the authority of the academy and the printed word of proper research to produce cultural truth and meaning. Instead, such sources assert the positive valuation of memory's transformative powers as a positive characteristic of a mutable live performance archive.

Archive of detritus

I would like to propose, speculatively, another possible alternative archive: namely, a theoretical archive of detritus. An archive of detritus would seek to mimic many of the positively valued characteristics of both the audience's memory of the performance and the disappearance of live performance.

To illustrate this it is worth thinking of the performances of Forced Entertainment, whose work will be considered in more detail in Chapter 4, and the manner in which it often highlights performance process through the accumulation of detritus on the stage. Many theatre productions clear-up as they go along: making tidy transitions from one act to another, the props from scene one quickly removed before the start of scene two. In contrast, the stage at the end of a Forced Entertainment production is often littered with traces of what has gone before, traces of the performance that was present but now has gone. Once noticed, this accumulation of performance detritus can be seen in many live performance productions. In, for example, Carles Santos' Latin opera *Ricardo i Elena* (2000), where the performers take their bows on a stage littered with pianos, picture frames, books and gigantic remote-control furniture, traces of the previous hour's events. For me the memory of the performance is contained in this final tableau, represented by remains, with the fragmented traces prompting fragmented memories. This is also experienced in Meg Stuart's dance work *appetite* (1998), which uses a slowly hardening clay floor to physically mark the passing of time on stage as the surface crumbles and becomes damaged as the dancers perform.[7] And once more in Wim Vandekeybus' *Scratching the Inner Fields* (2001), where the debris that remains on the stage – right down to a side-winder trail of sweat tracing a final movement of a dancer through scattered earth, sticks, and discarded clothing – are physical reminders of the moments that have passed before the audience.

Stage detritus presents an 'archive' able to create and re-create the multiple appearance of the performance. In the accumulation of these traces it is as if an immediate archive of the production is established, here is the shaky and incomplete evidence of what happened; these are archives that display their own randomness and selectiveness. These are archives that mirror the nature of the audience's memory of the production. These are remains that, uniquely, need archiving if they are not to disappear.

The image of stage detritus as archive is particularly suited for postmodern performance works that highlight their unstable and multiple natures, but also appropriate for the disappearing state of all live performance and of memory. The idea of detritus as archive is also not so far from the state of all archives: but the archive as detritus turns around the presumptions of neutral detachment, objectivity, fidelity, consistency and authenticity; instead claiming partiality, fluidity, randomness and memory. And having abandoned claims to accuracy and completeness, such an archive is able to present archival interpretations, proclamations and demonstrations; consciously and overtly performing what all archives are already enacting.[8]

4
Self-Representation

It would be difficult to quantify the precise relationship between the ideas, impulses and imagery of documentation explored in the previous two chapters and the more immediate world of performance production and reception. However, the passion in evidence within discourses of documentation suggests a significant connection with performances, practices and audiences. To work through some possible relationships, this chapter presents a detailed case illustration that traces the political and ideological relationship between performance practice, documentation, archives and the academy. This will be focused through the exploration of the documentary attitudes and activities of the British theatre ensemble Forced Entertainment, looking at how their practice is manifested in archival, academic and self-representational endeavour.[1]

In particular this chapter discusses how Forced Entertainment's work often appears to be accompanied by awareness of documentary possibilities, and at the same time how they also display somewhat contradictory (although never confused) attitudes to the activity of documentation. The integral, as opposed to merely pragmatic, importance of this contradiction will emerge through the course of the discussion. It is against the postmodern challenges explored in and enacted by the work of Forced Entertainment that it is possible to see how discourses of documentary urgency and practice appear at once to contradict, but at the same time also reaffirm, the cultural constitution and valuation of transience and disappearance.

Forced Entertainment

Formed in 1984, Forced Entertainment celebrated their twentieth anniversary in 2004. And in this context it is worth describing in a little detail some of the chronology between the company and my knowledge and experience of their work as a researcher, a student and as an audience member: a chronology that is relevant here not because it is unusual, or in any way special, but because I would suggest that it is fairly typical.

Founded just over twenty years ago (at the point of time of writing) Forced Entertainment has existed since before I was old enough to be interested in companies and productions such as theirs. Like many others, I first became aware of the company as a student, studying theatre at the University of Glasgow where we looked at the work of Forced Entertainment as part of a module on twentieth-century British theatre. Here I started studying the work of Forced Entertainment. This included going to see live productions as they appeared in Glasgow, and amongst other work I saw *Club of No Regrets* (1993) and *Speak Bitterness* (1994) at Glasgow's CCA as well as durational performances at the annual National Review of Live Art. Our studies, however, primarily focused on previous productions – particularly *200% and Bloody Thirsty* (1987), *Marina & Lee* (1991) and *Emanuelle Enchanted* (1992) – all of which neither my fellow students or I would ever see live. Instead, we studied these works through the writings and reflections of people who had seen them, in other words through academic discourses and documentations, and through watching video recordings of the productions (the experience of watching these videos is discussed in detail in Chapter 5).

For the university theatre student canonical performance-based work necessarily has to be taught in the absence of the live performance itself. Indeed, it is only through documentations of the performance – videos, photographs, archival documents, reviews, other writings and so forth – that it is possible for students to know, question or see performances of the past at all. In this instance the productions had occurred sometimes a mere year or two before, but were nonetheless as inaccessible to primary, live experience as if they had occurred decades in the past. Instead, as students we were invited to experience, reflect upon and analyse the works largely through the memories and primary experiences of others and our own experience of watching video recordings. Rarely do I remember this contradiction becoming a major topic of conversation and debate. Instead the video recordings were utilised as largely unproblematic notation of and even substitute for the absent live productions.

Since leaving university, my first-hand experience of Forced Entertainment productions has varied, dependent on a number of factors but most particularly on where they happen to be touring and where I happen to be living. In the absence of being able to catch shows live (the word 'catch', in the context of the disappearance of performance, being very telling) I have continued to rely on videos, written documentations and academic reflections. This kind of relationship with the work of a company or performance artist, with which one shares any kind of ongoing engagement, is in no way unusual. And although in no sense would I wish to lose my knowledge of Forced Entertainment productions (and others) that I have accessed only through documentations, it nonetheless strikes me as peculiar that a large proportion of my experience and knowledge of their work comes from videos and other documentations rather than from live performances. This is the pragmatic experience of the theoretical desire, explored in the previous chapters, of wanting to use documentation to 'see' live productions that would otherwise be unavailable to us; and yet at the same time the accompanying sense of unease that this is an unsatisfactory way by which to engage with the work on its own terms.

In an interview, Tim Etchells, artistic director of Forced Entertainment, quickly recognised this description of my personal chronology of experience and acknowledged that the work of the company is often first seen on video. Indeed, he is aware of people writing about the company whose *only* knowledge of the work is through documentary writings and video, something that Etchells describes as simply weird.[2] At the same time, however, it is the company that makes such videos available, that allows such weirdness to occur, and Etchells is aware of the value and need, yet accompanying discomfort, of putting such documentations out into the world. With this in mind, Etchells identifies Forced Entertainment as holding two parallel attitudes towards documentation, which he acknowledges as somewhat contradictory. These attitudes are: first, that which Etchells describes as 'extreme pragmatism, without ideology'; and second, a more nuanced approach that seeks to 'proliferate traces of work in a more artistic nature'.

Extreme pragmatism

An attitude of extreme pragmatism to documentation is often the immediate and inevitable response to the practical circumstances of making live performances in a world where, as discussed elsewhere, there is an instinctive impulse to try to record and hold onto things. For Etchells, documentation is the logical response to the need to make the work available to people who want to know about it and see it. And this pragmatism directs documentary approaches, media and techniques. Such pragmatic documentation, Etchells explains:

> Is done simply by asking what is a good solution to a particular set of problems you are facing. ... With good pictures your work will get into books, magazines and so on. Your work will get into different contexts.

This pragmatic attitude is present in Forced Entertainment's detailed and extensive company website (www.forced.co.uk), which includes large amounts of archival information – dates, descriptions, photographs, review material, interview, articles and so forth. The website meets many demands: from audiences wanting to know about tour schedules, promoters wanting information and contact details for the company, journalists wanting background information, and (not least) students wanting to research previous productions. This website is clearly functional, it is not about encountering the art of Forced Entertainment in any way, but more simply about communicating information. Anything more complex, more 'arty', would not be helpful, would be frustrating and would hinder the delivery of this pragmatic objective.

A similar attitude is evident in the company's production of video recordings of their work. For Etchells, 'video is there to promote, sell, to get the work seen. Without video it is hard to sell work to promoters and it makes it easier for journalists or academics to engage with work.' Pragmatic documentation, therefore, connects with and helps fulfil the most immediate pragmatism of all – economic survival. While Peggy Phelan laments, in many ways, how the production of performance documentations forces performance into a circulation of 'representations *of* representations' placing it within 'an economy of reproduction' (1993: 146), pragmatically this is often precisely the point. Documentations, particularly easily reproducible video and photography, allow performance to enter into markets of circulation and exchange that would otherwise be denied to them. To companies needing to engage with such markets this is vital. While Etchells agrees that video is a very unsatisfactory medium through which to approach their work as performance, it is for him the most practical response to the question of making the work more widely available. Such videos are useful for the information they can communicate and for their ability to satisfy basic but essential pragmatic purposes.[3]

What is clear in this attitude is that the production of pragmatic documentations is largely a response to external impulses and demands – whether they are from producers, for marketing purposes, or for study and the academy. Further, the ability for such documents to go where the live work cannot is vital to the reputation, visibility and general health of the company. In other words, the provision of such 'straight' documentations serves external needs rather than being driven by internal desires. For Etchells, this 'basic pragmatic approach is not to be confused with art making', with the two things unable to be done simultaneously.

However, as Etchells is also well aware, once work is distributed and made available through documentations, it is impossible to control how it is used and perceived. Problems arise if this extreme documentary pragmatism gets confused for something else, getting mistaken if not for art making itself then for being a complete and transparent representation, a replacement even, for the experience of the work itself. It is here that the desire to see work proliferated and disseminated more widely than through live performance alone, and more artistic attitudes to documentation, come together.

Artistic proliferations

Although the majority of the pragmatic documentations that Forced Entertainment produce are in response to external demands, this is not to say that there is no impulse to document, no urge to halt disappearance and preserve for posterity (no archive fever), emanating from the company itself. Rather, that these internally prompted documentations tend to take a very different form and serve different functions. Demonstrating once again the culturally recurring awareness and appreciation of the transient nature of live performance, and the desire to halt disappearance, Etchells points out that the fact that live work does not last prompts the question of how to proliferate its influences and traces in the world. 'Just to do live work' he states, 'would be the ephemeral of the ephemeral. ... The thing is to look for forms of objects that reflect the aesthetics and contingency of the live events.' Again documentation and disappearance go hand in hand.

This is what Etchells means by the production of more 'artistic' representations of the company's work, a category that he describes as including things such as their 10th anniversary reflections and performance/lecture *A Decade of Forced Entertainment* (1994), a piece which carries its own self-description:

> We wanted to look back at the decade 1984–1994 – the ten years in which we've been making our work, and we knew that this looking

back would have to include the things that hadn't happened as readily as those that had. We had in mind a map of the last ten years – a haunted map – a false map – and yet, in some ways, an accurate map. (Etchells, 1999: 29)[4]

Also included in this category of artistic representations, and demonstrating similarly allusive and non-literal attitudes towards documentation, are all of Forced Entertainment's written reflections, such as *Certain Fragments* (1999); some of the photographic work that emerged from performances, such as *Cardboard Sign Photographs* (1992); and most recently the interactive CD *Imaginary Evidence* (2003).[5] A central theme throughout these representations has been the desire to resist fixities, narratives, certainties and what Etchells describes as 'hideous linear discourses'. Recognising the accompanying need to provide empirical information – dates, times, facts, literal recordings – the self-conscious tangle of these representations actively kicks against the claims and pretensions of such straight documentations to tell the whole story.

These representations, therefore, seek a way of speaking about the work that is more akin to the aesthetics of the original piece. Often this involves mixing genres, using anecdote and conflating or expanding ideas and times and events to subvert conventional chronologies. The recurring metaphors for such representations are familiar, as they speak of fragments, traces and remnants. Such interests, such motifs, seek to provide a discourse that matches the postmodern interests and style of much of Forced Entertainment's work in performance, which is similarly involved in revealing its own fragmentations, constructions and contradictions. While it has become a truism to write of how *all* documentations of performance are only fragments of the thing itself, here this fragmentation is taken to extremes, as it becomes not merely the regrettable accident of documentary limitations, nor only the methodology, but also the ideology.

A good example of this is the publication of 'Notes and Documents' to *Emanuelle Enchanted*, which uses text and graphics alongside photographs by long-time Forced Entertainment collaborator Hugo Glendinning (Etchells and Lowdon, 1994). This imaginative rethinking of archival documentation does not present a clear or neutral documentation of the performance; it would be impossible to re-create *Emanuelle Enchanted* from this representation. Nor do Etchells and Lowdon attempt to interpret, evaluate or describe the performance. Instead, they accept the inevitable transformative effect of documentation and attempt to create a record that documents not the appearance of the event but instead represents something of the experience of the performance. As Nick Kaye comments on the documentation:

> [Etchells and Lowdon] use material derived from the performance to re-address concerns for excess information and incompletion. Rather than speculating upon the 'meaning' of *Emanuelle Enchanted* or recounting the mechanisms by which it operated, this presentation offers an experience analogous to that of a meeting with the event which preceded it. Calling on the 'fragmented/atomized' nature of the performance ... this 're-presentation' resists being read as a transparent record, but furthers the work's dissemination through a variety of forms. (1994: 6)

In his preface to a collection of performance documentations in *The New Theatre*, Michael Kirby suggests that the responsibility of documentation is to be objective: 'If it is a clear, accurate, objective recreation of the performance, the reader will respond to the documentation in much the same way as he would have responded to the performance' (1974b: i). Overthrowing this reliance on surface and neutrality, Etchells and Lowdon reject many of the impulses that direct the more pragmatic documentations. Instead, they present a representation that is far from clear, accurate or objective, which does not seek to re-create the performance but does manage to achieve the result of inspiring in the reader some of the experiences of the audience. The 'Notes and Documents' is an archive constructed not by recording the performance but by attempting to echo the memory of the performance. This is a fluid and transformative representation, an archive highlighting its own incompleteness, uneven qualities and the fabricated nature of its surface appearance. Unlike the pragmatic representations, these notes are not designed with any specific purposes in mind, but instead seek to proliferate the ideas of the performance in another form.

Moving several years forward and changing technologies quite dramatically, Forced Entertainment's CD *Imaginary Evidence* (2003) presents another meeting place between documentation and more nuanced representation. Described by the company as an interactive essay-come-archive, *Imaginary Evidence* continues many of the same impulses of the 'Notes and Documents', using the same metaphors of fragmentation, but now taking full advantage of the opportunities of the technology to further previous interests in mixing media, layering and crossing narratives and subverting easy and singular readings.

This ideology is made concrete in the CD's primary interface, where a brain-storm diagram presents a tangle of interlinking thematic possibilities that both point forwards and turn back on themselves as they present innumerable possible routes through which to access the material. One might, for example, travel from a point labelled 'Alcohol' to 'Failure' to 'Interruptions' to 'Sleep', but as users construct their own path through the material it is immediately apparent that there is no right or single route and indeed no single narrative in any one particular route. On the CD each of these thematic labels leads to a number of fragments – mainly video clips with voiceovers – with the same sets of material reclassified and reappearing under different themes. Moving through this material there is a recurring tension between free roaming inquisitiveness and a natural instinct to try and make sense of the material, to try and find answers and an overarching narrative or theme.

At this point there is something in the self-reflectiveness and nature of *Imaginary Evidence* that provides a slight moment of ideological unease. This is something that exists (although never front-stage, never worryingly so) in much of Forced Entertainment performance work, but which becomes much more apparent in their documentations. *Imaginary Evidence* includes a section about the CD itself, contained in a link entitled 'Ethos of the piece'. This communicates very much the attitude to the company's artistic documentations articulated by Tim Etchells and already discussed here:

> *Imaginary Evidence* is an interactive essay-come-archive, a machine for exploring ideas about performance in relation to the history and the work of Forced Entertainment. The collision of ideas, the juxtaposition of sense and non-sense, the drift from, around and to a point are its modus operandi. ... There is no correct sequence in which to explore *Imaginary Evidence*. Its composition is one of ideas, textures, facts and evidence that are repeatedly re-grouped, re-ordered, re-contextualised and, by implication, re-made. (2003)

This explicit ethos, and the nature of the CD, raises two interlinking responses. First, it is very difficult to imagine a method of reading the material in any other way than that prescribed in the ethos. As a result in a peculiar way *Imaginary Evidence*, which is ostensibly very much about openness and multiplicity, becomes itself a closed text. Second, and emerging from this first point, within *Imaginary Evidence* and throughout Forced Entertainment's self-representations (as opposed to their pragmatic documentations) it is clear that the motif of fragmentation becomes overwhelmingly pre-eminent. Again, ironically, fragmentation itself becomes the grand narrative, the overarching and stable meaning, at the same time that (ostensibly) such stable meanings are explicitly rejected. It is also the case that fragmentation exists as a prominent theme in Forced Entertainment's performance works themselves, although here a multiplicity of readings is also always possible, given the multiple layers and diversities of performance as a medium. In the self-representations, in contrast, fragmentation risks becoming a singular and excluding predominance. Here fragmentation almost becomes a mantra: of a postmodern world view as a whole, and of an attitude to documentation in particular. The potential implications of this situation, and the validity of this description, are worth exploring further, particularly by examining more closely the relationship between Forced Entertainment and the light industry of academic appraisal, analysis and the investing of authoritative meaning in performance.

Books, the academy and the assigning of meaning

For Etchells the production of artistic self-representations is motivated by the desire to contribute to the public, critical discourses – particularly academic – that circulate around and seek to define the work of the company. This is particularly the case with the written work, which 'attempts to map the work in a voice that comes from it rather than outside of it'. The 'Notes and Documents' of *Emanuelle Enchanted*, therefore, exist on paper, published in an academic journal (in 'book form' to return to the phrase discussed in Chapter 3). Significantly, this allows them to enter into and impact upon the academic debates and discourses that are constructed around the work of the company. Such discourses form a central part of how a work is categorised, defined and judged, and the act of self-representation seeks to establish a framework through which such analysis occurs.

An example of this relationship with the academy can be found in Kaye's book *Site-Specific Art*, which includes a documentation by Tim Etchells of Forced Entertainment's 1995/97 work *Nights in This City* (2000: 13–24). The documentation of this piece is fairly similar to the 'Notes and Documents' of *Emanuelle Enchanted*, which was also included in a volume edited by Kaye, describing itself as 'diverse letters and fragments relating to a performance now past' and including a number of self-consciously small and grainy black and white photographs. The text, presented as if extracted from the performance, also stresses the elusiveness of the work through a deliberate ambiguity and multiplicity:

> Ladies and gentlemen welcome to Rome ... this city is known to me for three things – the beer, the historical buildings and something else ... just there, behind these building, on the skyline you might just catch a glimpse of the leaning tower of Pisa ... and those of you who've been to Venice before will recognise the smell. ... (In Kaye, 2000: 14)

The possible readings that might be taken away from this documentation, however, are pre-empted by the context in which it is contained. Within Kaye's book the documentation, or account of the performance, is framed by an exploration of the work in relation to questions of place, space and site-specificity in performance. This framing does a great deal to focus and direct the meaning of the documentation to the reader. It becomes about such questions of site-specificity, rather than merely illustrative of it – almost as if the analysis is the cause of the work, rather than vice versa. Nor is this necessarily a question of reading one before the other, but instead is a marker of the power of the critique.

It is also intriguing how the letters in the documentation are addressed from Tim Etchells to 'My love', evoking doubt and speculation as to the addressee, who is deliberately unnamed and ambiguous. In many ways, however, it is tempting and not inappropriate to see Nick Kaye as the primary audience, for he is indeed in this context the patron, the commissioner, the intended reader and the addressee. In other words, Kaye (and the academy more generally) is the motivating force and purpose of this documentation.

Etchells' parallel commitment to but ambivalence towards academic reflections is also present in his attitude to a new book on the work of Forced Entertainment, *Not Even a Game Anymore* (Helmer and Malzacher, 2004), produced to mark the company's twentieth birthday. On one level Etchells is pleased by the existence of such a book, however, this is accompanied by a sense of the work and history of Forced Entertainment being taken apart and retold by other people. When invited to contribute to this book Etchells therefore responded with mixed feelings, but also a sense of the need to have a presence amongst these other voices. In particular there was a motivation to attempt to subvert the very activity and medium of the academic analysis, to question the appropriateness of producing such stable, refined considerations of the work of Forced Entertainment. Responding to these impulses Etchells' contribution, 'A Text on 20 Years with 66 Footnotes', echoes the nature of other self-representations discussed above, forming a tangle of narratives, with footnotes far longer than contents. Again this is a representation that is self-consciously and self-reflectively fragmentary.

An explicit riposte to the academic tone and very existence of the rest of the book, Etchells has an uncertain relationship with such academic considerations: aware of their importance and potentially framing, historical significance; of the longevity of such academic reflections; and also of their apparent authority. This awareness drives the acts of self-representation, motivating a desire to be involved, to be part of the discourse and to contribute (in a suitable manner) some form of self-definition. Etchells' own articulation of this experience echoes the discussion in Chapter 3 of the authority of the academy in constructing the performance canon: 'it is important to be in that dialogue ourselves, although it is clearly impossible to control. It is important because it can be influential in how you're seen and talked about. Books have power, they have a lasting and perceived authority.'

Indeed, the indirect but definite relationship between documentation, academic attention, reputation and continued economic survival is something that Etchells is very aware of.[6] He recognises the value of academic attention or of having a video presence in archives and libraries around the world, candidly admitting that one of the rewards of having videos of Forced Entertainment performances and rehearsals housed in the British Library is that it 'sounds good' and has a legitimising effect on perceptions of the company. Etchells is aware of the workings of prestige and reputation in relation to academic attention and respect, with indirect immediate impacts in terms of press attention and funding opportunities and long-term in the existence and acceptance of the company within performance history. To enter into the canon you need to play the game of collaborating with the academy and providing material to exist within archives. In this Forced Entertainment have been incredibly successful.

It is therefore possible to suggest that existence within documentation, and particularly circulation within academic representations and discourses, not only halts the disappearance of past productions (halting physical transience) but also ensures the existence of future productions (ensuring cultural endurance). Existence within archives and academic documentations provides performance with prestige and cachet, but also more fundamentally with legitimacy and a location within cultural/academic discourses. It demarcates the value of performance, which otherwise exists only in memory and elusive audience experiences.

At its most immediate level this is a question of resources and materials, on the existence of documents and representations that allow performances to exist within the canon, within research and within teaching. However, as Jill Davis suggests, is possible to argue that the academy does not simply document (or 'follow') performance practice but 'brings to bear the full panoply of modern cultural theory to locate, in fact to produce, the cultural meaning of these performance' (1999: 198). For Davis, the canon can be seen as limited to and containing only works that 'embody current theoretical issues'. In her case this is feminist theatre theory, although potentially we might see something similar occurring with Forced Entertainment – with the academy firmly locating the company within postmodern performance practices and, as a result, producing and determining the enduring cultural meaning and status of the performances. This does not only relate to the aesthetics and ideology of the performances, but also to that of the documentations.

Indeed, this articulation of fragmentation and the presentation of what might be considered self-consciously postmodern documentations can easily be seen as assuming a monotheistic status, becoming the canonical norm in performance documentation. In *Site-Specific Art*, Kaye includes documentations of work by companies such as Brith Gof, Station House Opera, Meredith Monk, alongside that by Forced Entertainment considered above, commenting that:

> one of the first observations one might make of each of these documentations is their sensitivity to their own limits, their willingness to concede the impossibility of reproducing the object towards which their statements, speculations, fragments, memories and evocations are aimed. (2000: 215)

Authored by the companies, the shared ideology of these documentations does indicate an intimate and non-accidental 'fit' with the interpretative, academic canon. In a mutually reinforcing relationship they are selected, or rather have selected each other, because they share the same concerns, interests, aesthetics and ideologies. The limitations of such a relationship is illustrated when Adrian Heathfield, for example, describes the *A Decade of...* self-representation as 'a fragmented narration of their performance practice', a response that merely echoes the work's own self-definitions. Somewhat similarly Kaye observes in the context of the 'Notes and Documents' of *Emanuelle Enchanted* that 'Like contemporary critical theory "performance art" has had much to say about the "trace"; the relic, recording or document that remains after the event' (1994: 6). The difficulty is that the repeated statement of this fact not only seems to contradict its own ethos but also results in commentators really saying (and being *able* to say) very little.

The worry, to reiterate, is not that the work of Forced Entertainment produces closed or singular readings, but rather that the discourses and documentations surrounding the work have the potential to do so, ironically enforcing singularity while espousing openness and multiplicity.

In their documentary activity, therefore, Etchells and Forced Entertainment are motivated by the desire to respond to the perceived authority of the academy and the lasting authority of the written word. The potential problem rests with the power that Etchells' own words have in directing (if not, as he acknowledges, controlling) readings of their work. This is particularly so given the position, discussed earlier, that the documentations hold of standing as entry points and gatekeepers to the work itself – as ephemeral performance we cannot see or know them except through these representations. The success of Forced Entertainment's acts of self-representation in this context is implicitly recognised by Judith Helmer and Florian Malzacher when they note that Etchells has marked his position as 'the first and most powerful interpreter of [the company's] own output' (2004: 12) – here the authority and endurance of the written word is joined by the authority of origin (of authorship). With the producer/author of the work also becoming the producer/author of representation, there exists a clear potential to close the discourse. This is combined with a kind of collusion between performer and academy, both of whom share the same motivation in this context, which results in the potential stability of a repeated articulation of fragmentation. Nor would I entirely exclude my own analysis here as exempt from this critique: either in terms of how, in interviewing Tim Etchells, I at the very least implicitly sought to obtain some authenticating primacy from having spoken to the artist himself, from having gone to the horse's mouth as it were; but also in terms of my being able to actively resist the invitation to read Forced Entertainment's work (or rather the self-representations of the work) in terms of qualities of fragmentation, trace and non-linearity.

What saves Forced Entertainment's self-representations from falling too entirely and restrictingly into this hole (if it is a hole) is that these artistic, fragmented representations are always accompanied by the pragmatic documentations, the straight archival recordings. This is made concrete on the *Imaginary Evidence* CD, where a clear, well-ordered, comprehensive and easily navigable archive accompanies the overtly tangled web of fragments. While the CD as a whole stresses postmodern credos of fragmentation, the archive contained within it implicitly yet confidently states its unreconstructed factual accuracy, completeness and authority. The grand narrative of fragmentation is undercut, in an ironic reversal of the common expectation, by the existence of the whole, of the well-written, factual, authoritative history. Rather than seeing the inevitability of fragmentation and constructedness behind the official history – the typical response to claims of documentary authority – instead we see the order, fact and authority behind the official fragmentation. In this manner, the self-declared contradiction in Forced Entertainment's documentary activity is vital.

Documentation and reputation

The relationship between Forced Entertainment and acts of documentation and self-representation takes the form of two interlocking attitudes and activities. The production and dissemination of pragmatic documentations of their work is about constructing a narrative of the *career* of Forced Entertainment – about direct relationships with curators, venues, festivals, funders, writers and audiences. The production, accessibility and quality of such documentations are immediately significant to the health and continued existence of the company. On the other hand, the aesthetic self-representations are about contributing to the narrative of the artistic *work*. Although not having such a direct and immediate impact, the production of more nuanced, artistic acts of self-representation also has a pragmatic motivation and purpose through their relationship to wider discourses of definition and critique and to the reputation of Forced Entertainment as a company.

There is an intersection and overlap between these two narratives: the same object, the same document can play a role in both. Nor is there a need to be puritanical in attitudes (something made concrete in the *Imaginary Evidence* CD, which contains both a deliberate fragmentation of themes and narratives and an easily navigable and comprehensive archive). And indeed, the central motivating factor for both activities of documentation – beyond any articulation of purpose – is always the same: the work does not last, does not speak for itself, and if its influence is to be extended beyond the moment of performance then it must be documented and spoken for in some form or another. These documentations can be pragmatic, essentially concerned with attempting to halt disappearance and communicate some trace of the performance; or they can attempt to, in some manner, continue the aesthetics of disappearance itself, communicating an aesthetics of trace, fragmentation, contingency and detritus.

One recurring theme in this book is the suggestion that it is necessary to study documentations not only in terms of looking at the traces and the content communicated about the performance, but also in terms of how the techniques and attitudes of representation articulate the values and meanings of the performance. In other words, focusing on *how* the performance is represented, as well as *what* (more literally) is recorded in any documentation. If individual acts of representation reflect upon the nature, form and meaning of the performance they are trying to document, then those of Forced Entertainment do mimic the elusive nature of the company's work.

Additionally, it is also possible to argue that attitudes towards documentation more generally form a representation of a company as a company. Tellingly, when speaking to Etchells he is able to talk lucidly about the attitudes of Forced Entertainment to documentation, both in terms of specifics and examples and in terms of a series of larger relationships and implications. The relationship between the company, their work and documentary representations and reflections is something they have considered in detail. Forced Entertainment as a company is very aware of and interested in their depiction and existence within documentations, archival representations and academic reflections. This documentary activity in and of itself, therefore, represents the nature of the company as a whole as a subtle and self-conscious entity, where strong artistic ideologies are accompanied by a pragmatic savviness that has allowed them to be consistently successful for over 20 years of making work. They show an awareness of the significance of documentation and a willingness and ability to play the game while at the same time trying to subvert and direct it to their own ends. The success of Forced Entertainment, in however we are going to judge and define the word success, is in part but not insignificantly the result of the success and activity of Forced Entertainment documentations.

Part II

5
Video Documents

Audio/visual recording technologies, particularly (in shorthand) video, offer the ultimate test case where questions of ethics, ideologies and practices in representing live performance are concerned. More than any other form or activity of representation, video is constructed within discourses of documentation and disappearance as at once both the saviour and the death of live performance – as at once something that will solve the 'problem' of documentation, and at the same time something that will potentially obscure and overwrite the original performance.[1] For while the ability of audio/visual technologies to record the appearance of performance in time and space prompts claims of faithful and mechanical documentation – neutral, objective, factual, complete – the very strengths of this claim simultaneously threaten the prized uniqueness of the live event, with the possibility that the mediatized reproduction becomes a show in its own right.

As a result, statements of loss, inadequacy and betrayal repeatedly underscore the relationship between video and live performance, with the cultural drive to document invariably accompanied by resistance and a positive valuation of disappearance and qualities of liveness. Elizabeth Zimmer is not untypical in describing how recording live performance on video felt like 'collaborating with the enemy':

> I've always been in the phalanx of the dance world that believes dance suffers mightily in the transfer from three dimensions to two. I was actively hostile to the notion of trying to can the live experience.
> (Mituma, Zimmer and Stieber, 2002: xv)

This sense of reluctance and resistance to the video recording of live work is frequently encountered, reiterating many of the disputes over the rights and wrongs of performance documentation encountered elsewhere in this book. Nicola Hodges, for example, writes of how performance artists 'feel a particular antipathy for videos that document their work as they feel, quite rightly, that a crucial ingredient of the performance is the fact that it is "live", an ingredient that is obviously lost on video' (1994: xvi). For those with an interest in representing live performance, therefore, both the attraction and repulsion of video resides in its status as a medium of performance in its own right – and in the transformation enacted therein from the live to the non-live. While it is not within the scope or ambition of this investigation to reopen long-running debates as to the relationship between the stage and the screen, or between the live and the non-live,[2] such questions will nonetheless be an implicit theme in a discussion where video splits those seeking preservation in documentation from those seeking another sort of preservation: the preservation of disappearance and the positive valuation of liveness and transience.

Accompanying such ideologically driven attitudes are the numerous possible practical and functional relationships between video and live performance, ranging in a continuum scale from the 'pure' documentation, to the for-the-camera stage re-performance, the studio reworking, the liberal adaptation and beyond to the original screen creation. Between these possibilities there exists inevitable slippage, as there has also been between the words documentation and representation throughout this book. The value of such categorisations, however, is that knowing the exact relationship between any particular live performance and its screen representation tells us much about what it is that we are watching and may crucially also change *how* we watch.

Across two chapters the following analysis explores examples located in broadly distinct and illustrative positions on this continuum of possible stage/screen relationships.[3] First, this chapter explores video documents of stage performances by Forced Entertainment (with the advantage of referring back illustratively to discussion of the company's other archival activities in Chapter 4). Chapter 6 then focuses on two examples of looser screen reworkings of stage performances – the first stage-to-film adaptations of theatre by the American Film Theatre; the second DV8's stage-to-television reworkings of their dance productions. Although originally conceived for film and television respectively, in both instances these screen representations are typically watched as video, and the exploration of the experience of video as video provides a key theme of analysis in these chapters. Drawing upon these case studies, these two chapters examine the ideology and status of screen representations of stage performances, considering both what these examples say about the specific performances they represent, about the medium of live performance and about the activity of 'videoing' more generally.

Video as a verb

There is a particular mobility in discussions about the use of media technologies, both in this specific context and more broadly. The difficulty is that arguments grounded upon contingent stages in technological development frequently become formulated as articulations of ontological truths. Writing on television in particular, Jane Feuer, for example, observes exactly this 'lack of historical consciousness' amongst commentators and a tendency 'to argue from an assumed "essence" of the medium' (1983: 12). However, when the technology shifts, and consequently cultural practices shift, such observations swiftly become redundant.

Here this observation has particular resonance, with this book being written at a time when a major UK high street store has announced that it will no longer stock VHS video recorders, which are giving way to recordable DVD with other digital recording technologies already on the horizon. This means that some elements of use currently associated with video will cease to be relevant to future activities. The exact nature of rewind and fast-forward is different in the digital rather than analogue experience; the familiar physical deterioration of the VHS tape will be replaced with different types of digital fragmentation; while improvements in visual quality mean that some of the texture of video will be irrevocably altered – for better and for simply different.

This does not mean that it is desirable to ignore the contingent experience of 'video' in our age: first, because to do so would be impossible; but also because to do so would be to elide the real experience of technology. Unquestionably, the dominant factor in the use and aesthetics of video has been the evolution of the technology – how it is employed following developments in its technological potential and possibilities. Nonetheless it does continue to be useful to think of video as a distinctive medium in its own right, with distinct and specific narratives of production and use that together coalesce to describe a multiple and unstable ideology of video as a medium. Additionally, this chapter explores how it is possible to see something more important than mere accident in the consequences of the technological limitations that have accompanied the use of video.

With such technological contingency in mind it becomes more accurate to think of 'video' as a verb, rather than a noun. Video does not name any particular thing, rather pointing towards an extremely broad and ever changing category of audio/visual recording technology. Video refers to a range of activities, a range of ways of recording, watching and reordering events and experiences. As an activity in relation to live performance, 'video' refers to the recording, watching and reordering of performances via audio/visual technology. It is for these reasons that video is often presented as both a medium and an activity. As Sean Cubitt writes, video *culture* is 'a set of relations around the uses of videotape, a set of practices and a set of possibilities concerning what these relations, uses and practices may become' (1991: 1).

The following two chapters, therefore, are primarily about video (videoing) as a cultural activity, and about the relationship between the impulse to video (the desire to document) and the available, accessible, changing technologies that allow this impulse to be fulfilled. More narrowly, they are about video as an activity in relation to the impulse to record, watch and reorder live performances, describing the relationship between live performance and its screen representations and exploring questions of ontology, ethics, ideology, practice and consumption. The questions explored are not so much about how (or why) to document live performance on video, but rather how to watch such video representations – asking what such experiences do and mean to our understandings and perceptions of theatre, of dance and of live performance.

The aesthetics of video

The promise that audio/visual recording technologies offer live performance is seductive, offering the potential to record, retain and preserve performance history; offering the potential to halt disappearance. This promise is eloquently revealed if we consider an imaginary archive of lost performances – echoing the frequent evocation of an imaginary museum of lost art works,[4] where the disappearance is the result of definite action (whether stolen, lost or destroyed), with an imaginary archive of performance where loss is more automatic, immediate and yet passive. Such an imaginary archive would naturally include everything from before the age of the film and video – as McAuley (1994: 185) asks, how much we would value a recording of any first performance of a Shakespeare play? – but becomes more tantalising if we consider the real possibility of video recordings of performances and performers that *might* have been captured on film, but were not. Shaemus O'Sheel, for one, laments the non-existence of film recordings of Isadora Duncan dancing, a sense of loss that could be salved by the discovery of a 'few reels of film by which the presence, the rhythm, the imperious gesture of Isadora could be at will evoked immediate and mobile – what a treasure that would be' (in Pearson, 2002: 110). In such a fantasy the promise, the allure of video is to let us see what might have been recorded, but which is lost and unavailable.

The allure of video is such that for some it provides the answer to all the calls for an accurate, vital and faithful method of documentation. Such belief often exhibits an instinctive (and hence often unconsidered) acceptance of video as a documentary medium. For film and video (and also still photography, considered in Part III, where this theme is developed in greater detail) possess the technological capacity to capture and fix an authentic and accurate reproduction of the world. Susan Sontag, for example, suggests that film is 'relatively speaking, a transparency, and it seems correct to say that one is seeing the event filmed' (1966: 25). It is this perception of transparency and directness, the ability to capture images that are mimetic to what we consider reality, which fuels much of the desire to video (as a verb) as a documentary act.

Additionally, in its narratives of production, video is frequently articulated as possessing several documentary advantages over other screen media such as film or television. While film production is expensive, large-scale, time-consuming and dominated by studios or producers, video recording is in contrast cheap, small-scale, immediate and individually accessible. Indeed, for many writers the possibility and ease for any individual to video positions the medium as fundamentally democratic, a radical incursion into the largely centralised control that is exercised in broadcast television and film production.

Video is indeed a hugely convenient and accessible medium: being portable and discrete (particularly today, less so in the early stages of the technology), affordable and involving relatively low production costs, a real-time activity that records sound and image simultaneously, allowing immediate playback without the need for development in a laboratory, easily and readily disseminated either on television monitors, through projection or on tape, in being relatively easy to use and in having a degree of permanence (although not absolute, as shall be discussed in a moment).

Together these elements produce a narrative of immediacy, with the production process being close, direct and instant, which in turn favours an aesthetics, an ideology even, of immediacy. This can be seen in the traditions and uses of video art, where qualities of the medium and perceptions of authenticity and immediacy are intimately connected. As Catherine Elwers writes, video art seems to bear the 'imprint of the artist's hand ... the artist's uniqueness embodied in his electronic mark making. More than with any other medium, we are momentarily convinced that we are witnessing the moment of creation' (2005: 12).

This perception of authenticity is affirmed by our perception of video as a somewhat raw, dirty, rough and unpremeditated medium. Indeed, particularly in the pre-digital era, the difficultly of editing videotape and the relative (and sometimes absolute) poor sound and picture quality are both the result of the immediacy of the technology (the result of the narrative of production) but also in turn bolster perceptions of the immediacy and authenticity of the end product. Cubitt, for example, echoes common sentiments in arguing that saturated colour, low resolution and degraded footage almost instinctively encourage viewers to read the image as a self-authenticating document. As a result, the use of low-band equipment can become a conscious aesthetic and political choice (1991: 139). In terms of the video recording of live performance these questions of the quality and readability of the recording – and (or perhaps *versus*) perceptions and understandings of authenticity and the status of the recording as a document – become vital.

These characteristics of video production have increasingly put the technology within the reach and budget of theatres, artists and companies, driven by the ability of video to record and thereby fix otherwise ephemeral live performance. Yet the archival status of video is by no means as clear-cut as this, with the medium possessing some particular characteristics that impact on the way in which it is produced, seen and used.

Ephemerality

That video provides a site for the permanent storage of live performance, that it halts disappearance and resists ephemerality, is a central mantra within discourses of documentation. Yet, as any archivist would immediately point out, there are serious limitations to the real permanence of video; particularly on magnetic tape, which typically has a lifespan of twenty years and can easily be degraded or destroyed through excessive heat, moisture, vibration and frequent playing. As Claire Daniel writes, those who espouse the use of video to address the disappearance and inaccessibility of live performance frequently elide the problem that every viewing, every use of the recording degrades its quality and longevity (2004: 85).

Indeed, the fragility of videotape, and particularly the fact that it degenerates with every viewing, has lead commentators to assert not only that video is ephemeral, but also that this ephemerality is not substantially different from that of the live performance. Auslander, for example, suggests that 'Both live performance and the performance of mediatization are predicated on disappearance' (1999: 45). Such observations of the ephemerality of mediatized performances apply across many different formats, including the disintegration of magnetic tape, the flammability of old film stock and the continual appearance/disappearance of the television broadcast flow and image. More prosaically, and less inherently, film and video may also be accidentally lost or deliberately destroyed. Indeed, in another parallel, Cubitt observes that as a result of all these factors 'it is impossible to see video art's history' as the video evidence itself is fading (1991: 86). This description echoes laments of the lost history of ephemeral live performance, with the similar result that the historiography of video art must reside in the memories of those who saw the pieces at the time.

Nonetheless, such ephemerality is surely of a different order and nature than the transience of live performance: magnetic tape is being replaced by digital and hard disc recordings (although by no means solving all archival difficulties in the process); film stock long ceased to be easily flammable; while the digital personal video recorder (PVR) records live television as we watch it, meaning that from an experiential perspective the ephemerality, the nowness, of even live broadcast television may come to be thought of as a technologically contingent moment in our cultural use of the medium. In contrast, the transience of live performance seems importantly different. Even considering magnetic videotape alone, in many ways the most fragile and ephemeral of the recording formats, there exists a fundamental perceptual and cultural difference in the nature of this ephemerality. For video gestures towards the condition of permanence, no matter how evanescent it is in its technological specificity. It may not be permanent, it may deteriorate, but both technologically as a product and culturally as an activity it *aspires* to permanence.

Douglas Rosenberg's comments here are typical, substitutable by similar observations from numerous others contributing to this debate, where knowledge of the fragility of video is always superseded by a discursive assertion of permanence in the face of contrasting live transience:

> Theatre offers no permanent storage for dance. After a performance one is left with the lingering yet ephemeral image of dance as it was in the theatre. Within the technology of video, the site for storage of the dance becomes the electronically encoded space of the videotape, allowing the audience to view it repeatedly as it is. (2002: 6)

This is also the instinct, for example, of Allegra Fuller Snyder, who asserts the documentary importance of video to dance, writing that 'Since video made it easier to capture movement in time and through space, the ephemeral aspect of dance was fast becoming less of an issue' (Johnson and Snyder, 1999: 7–8). Similarly, Michael Kirby writes that, 'The need for performance documentation lies in the nature of theatre itself. ... We have not yet reached the point where all – or even the most significant – theatrical presentations are recorded on film or videotape' (1974b: i). These declarations are unquestioned, unelaborated and unexplored: the implication being that to video would solve the documentary problem of theatre, if only all theatrical performances could be recorded – to video, therefore, is to resist disappearance.

Save as video

Part of the promise and allure of video, therefore, rests in perceptions of its transparency and neutrality (perceiving it not so much a medium in its own right, but instead an empty vessel waiting to be filled), its mechanical objectivity (recording whatever is put in front of it, capturing both audio and visual information), its permanence (making the performance available to replay and re-watch), but also in many ways something much simpler – its inexpense, availability and accessibility (with video often characterised as a fundamentally democratic medium).

Part of the promise of video also resides in various utilitarian functions, where, as with other forms of documentation, the video recording of live performance can serve many purposes, including: aiding the reproduction of past performances, facilitating the composition of new work, educational and research uses, employment in marketing and to project work out to mass media audiences. In any analysis of specific video recordings these overlapping functions and motivations must be understood and taken into consideration, as each vastly influences the nature and appearance of the recording and its relationship with the live performance event.

At the same time, however, it is telling that the same tape, the same recording, can shift from an initial marketing purpose to subsequent archival use; or be primarily created to aid rehearsal processes, but later be employed as a research tool. Indeed, whatever the original purpose of any particular video recording of a live performance, its eventual and overriding function is more broadly and simply that of documentation – of preserving and making present to see and know something that without being recorded would be inaccessible and unavailable. For each of these various functions also shares the important similarity of enabling the continued representation of live performance in its own absence. So Bob Lockyer states that video allows our national performance heritage to be 'preserved before it is too late' (2000: 41); for Judy Mituma videos 'make up a chronicle of dance history that will remain for future generations' (2002: xxxii); while, according to McAuley, 'film and video recordings at least enable us to counter the ephemerality of the theatrical event' (1986: 4). The basic function, beyond any subsequent use of video recordings, is to fulfil the fundamental purpose of halting disappearance and 'saving' performance from disappearance.

If the promise of video recordings of live performance is, in short, documentation, then the value of such documents can only partly be located in terms of their potential functions, but more fundamentally resides in their temporal endurance. What we use such document for, and subsequently how we watch them, is primarily impacted by our awareness that we can watch them at all. The function of video recordings is to enable people to see the performance, and, what is more, see it in its completeness, capturing both spatial and temporal, and audio and visual levels. The question, however, is precisely what kind of record of live performance is made available, and with what kind of relationship between an 'original' live performance and its screen documentation, representation or adaptation. The following section explores the narratives of production and use behind the first of these relationships, investigating ways of seeing and interpreting video *documents* of performance.

Video documents

Although unashamed proponents of adaptation abound, for many writing about the video recording of dance or theatre from a 'documentation' point of view, the principle response to what can be termed an adaptation/documentation dilemma is to insist categorically on video recordings being *of* live performances: that is of actual events, not representations constructed from specially made-for-camera performances. McAuley, on this point, insists that 'A recording made in real time during a "live" performance clearly has a different documentary status from a performance done in takes exclusively for the cameras' (1986: 20). Similarly, Annabelle Melzer writes:

> The only limitation upon the documenting process that seems to me indisputable alongside varying intentions, is that the film be shot in performance, with or without an audience, *in the original setting* – that

is, not shifted to a studio. The shift to a studio space seems to me the critical dividing line between documentation and adaptation. Transferred from its original space, the performance must become, in its filmed version, an adaptation. (1995a: 152)

The motivation for such distinctions between adaptations and documentations resides in attempts to ensure authenticity and thereby the continued utility of video as a transparency and way of seeing the performance by proxy. For the less 'faithful' a film/video representation of a live performance the less 'useful' it is as a document of that performance. With video documents intention and attitude is crucial: the ambition being to capture and archive the live performance, rather than create any kind of alternative performance experience. For each reworking of the stage performance for the camera complicates the relationship between the screen artefact and the live performance it represents.

At the same time, however, in adapting a performance for the screen each step made away from direct fidelity is typically taken according to what will make a more successful video performance in its own right, or according to choices directed by 'what the camera can conveniently record rather than a reflection of the theatrical functioning of the piece' (McAuley 1986: 9). In this manner, instinctively and almost inevitably, performance is transformed through the act of recording. As Rosenberg suggests, the video camera can have a coercive affect on what it films as 'the camera tends to exert a sort of authority that shapes a situation it intended to simply reveal or fix' (2002).

The recurring dilemma of video recording live performance, therefore, is whether the practices of representation are determined by what is considered valuable and worth documenting, or are instead defined by what is documentable. The more faithful the video representation, and the less it adapts the performance for the new medium, the less watchable it becomes as an artefact in its own right. Consequently, in terms of video documents of live performance, two key and overlapping points predominate: first, everything about such recordings asserts their secondary status as copy of a live original; second, the watchability and aesthetic value of the video record is less important than its factual content and very existence.

The video recording in this context becomes an archival document, a status most clearly demonstrated in the use of video as a research tool and its valuation amongst performance scholars. Indeed, the ability of video to bring performance into universities, to make it study-able is widely proclaimed, with McAuley suggesting that video technology means that performance 'can finally become a central focus of analysis' for theatre scholars (1986: 2). Similarly, Sherril Dodds writes that video is 'an essential tool for studying dance irrespective of its ephemeral nature or geographical locality' (2001: xi), while Varney and Fensham assert the absolute value of video to the researcher:

> Video is a necessary and unnecessarily maligned aid to research; without it, performance disappears and we lose our history and our capacity to think through performance. (2000: 89)

The implication is that through video archives and through the medium of video the study of performance (and almost by implication its very cultural existence, as discussed in Chapter 3) retains a future that is otherwise unavailable.

The construction of such archival video recordings of live performances has become very much the norm, the default practice of theatres, production companies and also museums and archival institutions. Indeed, in her research into current practices in video recording, both in the United Kingdom and internationally, Claire Daniel describes an unsystematic but fairly dense and overlapping mesh of national recording and collecting institutions – notably in the United Kingdom, the British Sound Library and the National Video Archive of Performance at the Theatre Museum – alongside activity by individual theatres and production companies. While not all venues or companies operate an in-house, archival recording policy, those aspiring to produce work that they believe 'will be of particular interest to future generations' generally do (2004: 18–34). This motivation for activity is another marker of the wider archival drive identified in previous chapters, with documentation and archival recording a gesture towards the future employed as a marker of enduring value.

Within the category constructed here of 'video document' a range of different recording practices coalesce, with variations including numbers and location of cameras, different kinds of camera operation (fixed, mobile, operated, automated), and different attitudes and practices in editing. Across a range of practices, however, the overriding conceptualisation of the video as document prevails, recording the performance as seen live (frequently during an actual performance and under the available lighting conditions) and capturing all of the performance as seen by an audience. The Theatre Museum, for example, uses between one and four cameras (depending on the size and nature of the production) with the different feeds edited together. However, Daniel notes that beyond the specialist recording activity of such institutions, 'almost all theatres and theatre companies make their recordings using a single fixed camera' (2004: 50).

It is such fixed camera video recordings, still typically recorded on VHS, which form the archetypal video document of live performance. Part of the reason for their ubiquity and quasi-iconic status is that they utilise many of the particular narratives of video production described earlier – being largely unedited, cheaply produced and bearing a direct and immediate relationship with their subject. The appearance and failings of such recordings are also familiar – the picture quality is invariably poor, sound quality worse, colour either saturated or washed out, detail lost in low resolution, while the static camera renders even the most dynamic performance slow and stilted – again adhering to the characteristic aesthetics of video as a medium of raw and dirty authenticity. The following discussion explores video documents further, looking at the implicit demands they make on viewers and how they invite particular reading strategies and relationships. While in terms of video as a medium (as a noun) many of these characteristics may be contingent accidents, erased by technological improvements, these single camera recordings crystallise video as an activity (as a verb), as an instinct and as a culture.

Forced Entertainment on screen

The video documents of live performances explored here are all recordings of Forced Entertainment productions. In this context it is worth recalling the discussion in Chapter 4 of artistic director Tim Etchells' complex and self-aware attitudes towards the production of various kinds of documentations and representations of the company's work, including academic reflections, articles, books, photographs and interactive CDs. Against this background, the company pursues a largely pragmatic and deliberately unsophisticated approach to their screen representations, producing video documents as archival documents in the manner described above.

As illustration of this, it is possible to look at almost any video of a Forced Entertainment production, which are usually recorded during live performances, using available lighting, capturing the performance as performed on stage and employing very simple technological and editorial approaches. *Marina and Lee* (recorded Sheffield 1991), for example, is recorded using a single camera, located at the back of the auditorium and covering the entire theatrical event: from the audience taking their seats, to the applause at the end of the production. Although fixed in position the camera is operated, able to pan and zoom (albeit in a fairly jerky and awkward manner), to follow movement and provide some relative close-ups. The screen picture, especially when reproduced on a small monitor in viewing room of a library or archive, is indistinct, unclear, blurred, washed out; the performers are distant and their voices faint. This particular experience is exacerbated by the worn nature of the videotape (described earlier as itself part of the particular aesthetics of video and immediately very familiar, particularly with library copies), which can possess a scratched and degraded quality. It is a video that could never be watched for pleasure, but only for some other utilitarian function.

Other Forced Entertainment productions, including *200% and Bloody Thirsty* (recorded Nottingham 1988) and *Club of No Regrets* (recorded London 1993), offer slightly more sophisticated recordings. Both of these examples utilise three cameras, with a combination of long-shots and close-ups edited together to form a coherent artefact. In each of these recordings there is nothing surprising in the shot selection, with close-ups used almost in sequence and by rote: providing first context, then detail; introducing each character in turn; focusing in during monologues, widening out during larger stage action. These recordings seek to assert an aesthetic of neutrality – of commonsense, of document-ness – and although both these examples are more 'watchable' than the single camera recording, they are in no sense gripping or engaging. If one of the key relationships that Forced Entertainment seeks with their audiences is that of implicated witness rather than detached audience, then by contrast the video viewer remains alienated and excluded.

In conversation, Etchells acknowledges the limitations of these video recordings, very much echoing the description here when he observes that 'the work doesn't work on video [and] is hard going'.[5] This brute statement is both true and problematic. If this is the case then why release the works on video, and certainly why release them in this very problematic form? Particularly when, as Etchells is aware, for some audiences Forced Entertainment's work is often first seen on video and sometimes even *only* known on video. The answer lies in the utility and explicit *usefulness* of the video document, with Etchells describing various overlapping motivations for producing the video recordings, including using it to sell work to promoters and making it easier for journalists and academics to engage with the work. More broadly he is aware of how the existence of the works on video allow them to gain a far larger and wider dissemination than if they were limited to the live performance.

This utilitarian approach directs not just the existence and dissemination of the video recordings, but also their form. As discussed in the previous chapter, Forced Entertainment have produced many sophisticated reflections and representations of their work, but for Etchells the status of video as an audio/visual performance medium in its own right makes it trickier and more problematic as a form for such artistic representations. More immediately, it is worth asking what the use or purpose of such a video be: any attempt to make a recording more watchable would obscure the informational, pragmatic qualities of the recording. For Etchells the one or two camera video, using simple editing and few close-ups is a 'purer documentation' that 'speaks better of the work than a multi-edited production'. The faith here rests on the notion that in being as bare and direct (as unedited and unarty) as possible the video document possesses a kind of neutrality; that it is a transparency that viewers can see through to the performance beyond. Indeed, the gritty, dirty nature of the recording almost seems to serve as a marker of authenticity: the unglossy nature of the rendering an assertion of its faithfulness. This, however, is not the case, as even the most unsophisticated rendering of performance on video still renders the performance as video. As the medium of the experience changes so does our way of viewing and watching also have to change.

The manner in which these video documents speak of the work, therefore, remains compromised, as can be seen in the experience of watching other recordings of Forced Entertainment productions, such as *Some Confusions in the Law about Love* (recorded Sheffield 1991, again recorded by a single camera). Here the performers' faces are often pixilatted into a featureless pink blur; the camera frequently losing focus and contrast, the screen image dissolving into a drunken starry sky before recapturing, regaining its sharpness. In this recording the unsophisticated camera equipment had problems coping with the available lighting levels and the degrees of contrast, with the result a recording where the technology itself seems to be continually and explicitly grappling (and failing) with the problem of representing a different medium. It is intriguing to wonder exactly what kind of attention should (and do) we pay to this kind of limitation and distortion within the video document. (Intriguingly appropriate in this context given Forced Entertainment's own interest in the aesthetics of failure.) Is it feeble and improper to notice such defections; should they be ignored as we make 'allowances' for the deficiencies of the video rendering? Alternatively, is there potential to elaborate a semiology of such accidental special effects? For example, one of my experiences of this is how the inability of video to cope with high levels of contrast frequently results, with white skin, in a screen-pixilated blur obscuring scenes of nudity, almost as if the recording is enacting its own form of technological censorship. Either way, there is again little pleasure in the experience of watching, merely a sense of utilitarian purpose accompanied by a slight confusion about how to watch the recording at all.

To video culture

As a medium, video permits several distinct activities or interventions on the part of viewers that impacts on the way in which it is watched and used. These activities are familiar, including the ability to pause, fast-forward and rewind (including the ability to do all 'in vision') and to replay. Each of these allows viewers to intervene into the recording and to construct their own performance through a crude form of editing. So, for example, Julian Wood describes how young boys fast-forward through the dialogue and exposition scenes of horror movies, in order to construct their own, 'stronger' experience of the video (1993). Such possibilities impose an instability on the recording, asserting that it does not only have to be seen one way, or only once, and provide viewers with a limited form of power and control.

Additionally, the status of video as a recording medium provides video viewers with the overriding ability to 'timeshift'. A video can be watched at any time, fixed neither by the scheduling of broadcast media, by the programming of cinema screenings or indeed by the temporal and spatial sitedness of live performance.

Once opened to timeshifting, and once opened to activities of pause, fast-forward and replay, a recording on video is opened to a much more dispersed and fragmented relationship between performance and viewer than is the case with other audio/visual media or indeed live performance. While video as a recording technology remains fundamentally linear, as an activity and culture to video (as a verb) allows multiple fragmented and diverse experiences. With video the relationship between performance and audience reaches its most exploded and dislocated position, with temporal and spatial simultaneity shattered and with potentially no continuity or cohesion within the audience experience either in single experiences or subsequent experiences. Not least among the dislocating possibilities of video is its impact upon our understandings of liveness. If liveness is constructed as absolute temporal and spatial simultaneity (here and now), then video represents the instable opposite extreme. If live performance is determined by the co-presence of audience and performers, then video is delimited by a fundamental absence.

In terms of the cultural use of video, one consequence of this dislocation is a characteristic relationship of distraction and inattention between viewers and screen performance. Many elements of the live performance experience – including the ritual and social nature of the event, the collective attention of the audience and the co-presence of actor and audience – encourage or determine what can be described as a heightened level of attention and engagement on the part of the audience. Elinor Fuchs, for example, writes of a 'circle of heightened awareness flowing from actor to spectator and back' (1985: 163), Martyn Evans stresses the effect of 'listening under circumstances of heightened tension in the highly charged atmosphere of the recital room' (1990: 9), while theatre designer Iain Mackintosh describes how the intention of theatre architecture is to provide 'a channel for energy' to be carried from performer to audience and back to performer (1993: 172).

Maintaining this description of 'heightened space' of live performance, sociologists Nicholas Abercrombie and Brian Longhurst also contrast the high attention rates of theatre performances to the low attention rates of television and home-video viewers (1998: 40–3). Indeed, distraction can be termed as central to the experience of watching video, as the specificities of the medium (the ability to pause and replay) and the circumstances of the viewing (typically at home) mean that there is little sense of occasion and seldom urgency to the watching experience. This is not to say that the video viewer is always distracted, or conversely that they cannot be gripped and enthralled (with the viewing of rare archival footage of live performance providing one potential example of this), but rather that the medium in its own right does not demand concentration or assert specialness. There is no need to watch *now*, one can always watch later. As Michael Kustow, commissioning editor for the arts on Channel 4 from 1982–92, puts it:

> When you're in the same space with performers you'll give them more time and things can develop more slowly, because you can see the whole thing and you're sharing. When you're watching the television ... It's not that there's a three minute attention span, I don't even think that's true: it's that because there isn't any presence, things have to be stated more economically, more swiftly. (In Marigny and Newman, 1993: 96)[6]

For many artists working both live and on screen, this different level of audience attention and engagement means that there are things that work live but that do not work on screen, not least in terms of the pacing and length of the performance as Kustow notes. For dance filmmaker Eiko Otake this different level of attention must be acknowledged: 'we recognise that on screen we, as performers, have either to compete for viewers' attention or to be patient with their inattention' (2002: 87). Of these two possibilities the former is only available to those willing to adapt the performance for the camera, with the latter the only option for the video document – although it is also interesting to consider the latter possibility as both the more ideologically challenging and the more apposite to the particularities of video as a medium.

In the video document of *Some Confusions in the Law about Love*, for example, there is a long section at the beginning of the recording of the audience arriving. During this section, which lasts approximately eight minutes, the camera is fixed on the indistinct and empty stage. Like me, most viewers of the video, naturally, fast-forward quickly through this section until they reach the beginnings of the performance 'proper'. Unable to get anything of the mood, the buzz, the sense of anticipation that occupies the live audience, this passage is one of simple tedium. During the live performance, the length of time would be unremarkable, the audience being within the sphere of darkness and anticipation themselves (rather than excluded from it by the screen) and drawn into a heightened level of attention and expectation. On screen, in contrast, this opening is awkward and feels far longer than it actually is, marking out differences in reception modes and experiences between the two media. A mundane example perhaps, but watching video recordings for this book I soon noticed that I was failing to watch any of them straight through, instead always making use of fast-forward button to speed up the recording, edit out 'tedious' sections and skim over the difficulties of the watching experience. At the same time, however, I was also making use of the ability to replay and freeze-frame sections of the tape, and in doing so was already imposing the particular aesthetics and experience of video onto the representation of the performance.

Through such use of fast-forward, freeze-frame and rewind, viewers of raw video documents of live performance become editors in their own right, each time constructing a new text apposite for the particular uses and purposes of their viewing. Such usage takes full account of the specificity of video as a medium and as an activity, rather than employing it as a transparency to distribute other content originating on other medium. Such usage is also apposite given that the function of video documents such as that of Forced Entertainment cannot be entertainment, cannot be the replication of the experience of the original performance and therefore has to reside elsewhere – primarily in utilitarian notion of video as a study tool and archival document.

On video literacy

This description of the utility of video documents at the very least implicitly recognises that the recordings convey little of the impact of the live event, little of the dynamism of the performance, the emotion, or the charged nature of the audience experience. The suggestion, however, is that *something* of the performance is available to us, particularly through careful study, through attention to detail, through the (absolutely) hard work of watching. As Virginia Brooks writes

> Although these films may not be visually interesting in their own right, they are valuable tools, for they permit careful analysis and preserve one of the means by which a work may be reconstructed or restaged with accuracy. (2002: 48)

In a form of exegesis, the suggestion is that viewers need (and even have an obligation and responsibility) to unpack the video document and learn to *read* the screen. Assertions of this need are common, with McAuley arguing for the need for 'reading' skills when looking at video documents, where a different viewing technique is required from both the experience of the theatre spectator and that of television viewer (1994: 184). Dave Allen similarly writes of how dance recordings may seem merely quaint or amusing 'unless we learn to contextualise and *read* the work in ways that make it productively accessible' (1993: 19). Such arguments state that we must make allowances for the deficiencies of the recording and that we must work to interpret the information contained within the documents. This is very much a description of archival endeavour and textual scrutiny: opacity forces us to strain to see, but also assumes our engaged and specialist interest, rather than that of the distracted video viewer. What the video document certainly does do is turn the event into archive; this archival quality not only means that we have to study it, but also attempts to make us want to study it – something that is not incompatible with the experience of fast-forwarding through the recordings, which can be thought of as skim reading for salient information, rather than watching raptly for the experience.

The need to work hard to read these video documents describes the need for a kind of literacy that is distinctly different from televisual literacy, from the screen codes that are familiar and largely instinctive to our everyday watching experiences. Indeed, not only do such video documents not follow televisual grammars,[7] but also often seem to actively resist them. The video recordings contain all those elements of theatre that translate least well to the screen – such as slowness and also staginess – and utilise few of the benefits of screen media.

Here it is worth thinking of Marco de Marinis' assertion, frequently echoed elsewhere, that we must not allow the recording of a live performance to become a 'surrogate' show, supplanting and replacing our understanding of the original live performance (1985: 386). Possible methods he proposes to ensure the secondary condition of the video documentation revolve around maintaining the visibility of the recording medium. In particular, de Marinis advocates the construction of partial and intentionally awkward recordings that deliberately resist attempts to be read as television. In this manner, by making the recording medium as evident as possible, viewers will never be able to forget that they are watching a representation of another event and not experiencing a work in its own right. This ambition removes the video document from attempts to assert authoritative objectivity and provides it with a very different role than that suggested by the employment of the medium as a transparency. By deliberately seeking to bring the limitations of the medium and the fact of mediation to the forefront of the representation, the absent live performance becomes, ironically, more visible.

Almost by default, this is already the status of these awkward and inaccessible video documents of live performance, where viewers are never able to forget that they are watching a video, a copy, an ill-fitting rendering of something else. There is an aesthetics of CCTV here. Video as a form of evidence, that records the time-based facts of the event, that needs to be interpreted, scrutinised and enhanced not only for the purposes of analysis but also simply to be able to see what is there at all.[8] They are recordings that are not easy to watch, but which make the compromised and observational nature of the watching explicit.

However, there is a tension within any ambition to resist the instinct, the necessity even, to watch video *as* video: especially as the very fact of watching, the very possibility that the performance can be watched at all, already embeds the experience within the particular characteristics of video as a medium and activity. Even more specifically than conforming to broad televisual grammars, when we watch video as video, we do so according to the particular aesthetics and cultural practices that are embedded within the medium and within our use of the medium.[9] The challenge of reading against this video practice is not to be underestimated. The result is that watching the video document becomes a kind of hybrid experience upon and within which we begin to construct our own narratives layered somewhere between the video medium, the live performance medium and the activity of watching – of videoing – itself.

6
Screen Reworkings

Any discussion of the representation of stage on screen runs the risk of perpetuating a hierarchical binary relationship that constructs the live performance as the original and the screen presentation as its copy. This chronological hierarchy also often assumes implicit evaluative judgements: loading the 'original' with a primacy that resides in its authenticity, with the representation by contrast becoming 'merely' a copy, inauthentic, unoriginal and secondary.[1] Such a binary can, additionally, lead to assertions that the live is intrinsically superior to the non-live, in part because of various perceptions of authenticity.

This constructed and value loaded relationship is something that poststructuralist critics have sought to deconstruct, not least because it is evident that the movement between live and non-live performance is not only in one direction. While these chapters focus on the non-live representation of live performance, and while theatre has long provided source material for film and television, there are many examples of works first conceived for the screen then being adapted for the stage.[2]

However, the relationship is much more nuanced than merely identifying whether the stage or screen version came first in any one particular instance. The presence of screen stars in stage productions; the referencing of screen genres; the use on stage of projected images, television monitors, surtitles, microphones, synthesised sound effects and so on, are all further examples of the quoting and invoking of screen codes, literacy and expectations in live performance. More specifically, video maker Douglas Rosenberg suggests that live dance has increasingly begun to mimic filmed dance, with the result a kind of 'corporeal reproduction of technologically mediated performance' (2001), while Alan Durant suggests that with live performances of music, 'What is represented on stage can become primarily a version of something that can be heard on disc and is for sale' (1984: 111). Within such interlocking layers of mediation there exists an erosion of concepts of original and copy, a description echoing Jean Baudrillard's assertion that the very opposition between original and copy has been eroded in an age of simulacra.

This theme was present at an implicit level throughout the previous chapter, particularly in the exploration of how the representation of one medium of performance (the live) in the form of another (video and the non-live) impacts upon the spectator's experiences, cultural practices and viewing strategies. This chapter develops the theme further, through the exploration of much looser screen reworkings of live performances: first stage-to-film representations of theatre; later stage-to-television adaptations of dance. In both instances, the screen performances retain a sense of being representations of a live original (they are not entirely new screen creations in their own right) and this chapter explores how these residues and echoes of an other and absent performance impact on how we watch and respond to such video representations.

Theatrofilm

Amongst the various utilitarian purposes motivating the production of audio/visual representations of live performance the archival, documentary drive is almost certainly paramount. If this archival drive is partly about 'saving' the work from disappearance and for future generations of audiences, it is also partly about the desire to reach larger audiences (not only in the future but in the here and now) and disseminate a work beyond the scope of its theatre manifestation.[3] As Virginia Brooks writes, 'there are many people who cannot attend a live performance of dance, and they deserve a chance to see performance on a screen that truly convey what the dances look and feel like in the theatre' (2002: 59).

However, if the ambition is to allow wider audiences to see a performance then the dark, slow, unclear and undynamic video documents, as exemplified in Chapter 5 by recordings of Forced Entertainment productions, are clearly not appropriate. Although there does exist the possibility of a certain fetishism of the degraded image as a marker of authenticity, it would almost certainly be unrewarding to pursue this argument into a fully fledged campaign that positioned magnetic tape as a new vinyl, possessed with soul and heart in contrast to the empty perfection of digital recording techniques. As Elizabeth Zimmer suggests, 'prime-time quality recordings' are essential in potentially reaching new audiences (Mituma, Zimmer and Stieber, 2002: xvi). The subject here, therefore, is not straightforwardly the matter of adapting material from one medium to another (from novels to film, from theatre to television and so on), but rather the ambition of employing the reproducibility of screen media to project theatre out to larger audiences and yet maintain, as Brooks demands, the distinct nature and character of the original stage performance. In other words, the ambition is to somehow convey or transmit the live medium of theatre through non-live screen media.

Two instances of this ambition, the non-live dissemination of live theatre, are worth briefly exploring here as illustration of this somewhat problematic agenda. The first example is of what Daniel Rosenthanl describes as a version of *Hamlet* (dir. Bill Colleran, 1964) 'unique amongst Shakespeare films' in how it sought to screen disseminate a stage production. The film was of a New York stage production, directed by John Gielgud with Richard Burton as Hamlet:

> Three performances were recorded through the 'revolutionary' Electronovision process, which used small, electronic cameras that could deliver adequate picture quality using only available stage lighting. Five cameras were used and their positions switched nightly so that the edited 'theatrofilm', as *Hamlet* was billed, could combine fifteen viewpoints. (Rosenthanl, 2000: 28)

Beyond this the extent to which the production was adapted for the screen was minimal, as according to Burton himself, 'none of the actors made any concessions to this new process. In other words we didn't tone it down in order to seem like film actors' with the results described by *Variety* as 'distressingly dark'. Also indicative of this project's compromised status was Burton's insistence that, with the motivation of protecting audiences for future theatre performances, the film only be shown for two days and that afterwards all prints be destroyed. Perhaps partly because of this, and the urgency and quasi-liveness invoked by such a short and prescribed distribution, the film was a huge success – seen by an estimated 5 million people across 970 American cinemas. In a final twist, Rosenthanl describes how 'three rusty film cans' were rediscovered in 1991, with the cultural value of the film now having been transformed from that of contemporary entertainment into something intrinsically archival.

If 'theatrofilm' represents one clumsy neologism for an awkward attempt at a new form, than the term 'electronic theatre' coined in Canada in the late 1980s is another. According to Richard Kirkley, the intention was to establish 'electronic tours' of theatre productions as part of a scheme to enable increased access to the theatre and as a viable alternative to the high costs of physical touring. The proposal was that venues around Canada would host high-definition large-screen television performances on a simultaneous feed from the live performance happening elsewhere in the country. A report on the possibilities of such a nationwide programme, *Accent on Access*, predicted that, 'When the curtain rises for Ottawa audiences, it will be rising at the same moment in cultural centres across the country for the benefit of all Canadians', although the idea of continuous transmission of live performances using multiple cameras had to be rejected because of technical difficulties (in Kirkley, 1990: 4). As Kirkley points out, the project also raised important questions over the different experience of each medium, including the reciprocity between live performer and present audience and contrasting imaginative engagement with qualities of realism (1990: 5–6).

Indeed, the question that these film/theatre possibilities raise is of the precise relationship between form and medium: or crudely, is 'theatre' defined by its medium, which is live performance, or by its form, which is drama? This is a question that seems more urgent and necessary in relation to theatre than it does to other forms of live performance, such as dance – as will be discussed later in this chapter – for the difficulty is that while drama may be broadly the fabric of both film *and* theatre, the particular rendering and experience of this content in each medium exposes the differences between the two forms. Not least in how to describe a screen performance or production as 'theatrical' or 'stagy' is invariably to insult it.

With regards to the relationship between live and non-live drama such questions are particularly loaded, as can be seen linguistically in the way that mediatized drama is no longer theatre (the word reserved for the live performance) but instead becomes film or television (words also defined by their medium). Consequently, once represented on screen, it is possible that audiences no longer see theatre as theatre but as video, film or television. The screen representation not only becomes a surrogate show, but in fact the only show that we are able to see (literally, given the emphemerality and unavailability of the live performance). The problem remains, however, of how to watch this peculiar kind of show, located as it is somewhere between our experiences and expectations of both theatre and film or television.

American Film Theatre

A good example through which to illustratively explore these kinds of questions is the work of the American Film Theatre (AFT) in the early 1970s. As a case study, this example demonstrates an attempt to use screen media to bring larger audiences to theatre plays that bears some similarities, particularly in terms of ambition, to Burton's theatrofilm or the Canadian electronic tours. More so than these instances, however, the work of the AFT also sought to 'open up' and adapt a stage performance for the screen while at the same time maintaining an increasingly problematic relationship with a prior theatrical 'original' and retaining a quality of being theatre on film. The resulting AFT productions are also interesting in that when first conceived as a cinema-based project in the 1970s they largely failed to gain significant audiences, but have more recently been given a second chance to succeed through their re-release on DVD – thereby repositioning these film/theatre productions within video culture and use.

For two years in the early 1970s, the AFT sought, in its own description, to bring together the best of film and theatre, producing film versions of classic twentieth-century play-texts, featuring 'impeccable performers and wonderfully established directors' (Landau, 2004). Under the direction of Ely Landau, the AFT produced a total of fourteen films in 1972–73, including Lee Marvin in *The Iceman Cometh*, *The Three Sisters* directed by Laurence Olivier, Jean Genet's *The Maids* with Glenda Jackson and Susannah York, Edward Albee's *A Delicate Balance* with Katherine Hepburn and Paul Scofield and Harold Pinter's *The Homecoming* (which will be examined in detail in a moment).

Across this range of productions various possible relationships between theatre and the screen were utilised: sometimes merely employing talent drawn from the theatre; sometimes 'opening up' a stage production for the camera; sometimes presenting a more or less literal record of an actual stage production (although never simply recording an actual performance). Some of the films, therefore, include actors who never performed the role on stage; others were produced even when there was no recent or contemporary stage version to adapt/record. All the productions, however, remain representations of theatre in that in each case the explicit ambition was to present *theatre*, but through the medium of film. For, rather than adapt the plays for the screen or find their cinematic equivalent, the AFT started from the premise that the plays were theatre plays. As Richard Peña puts it, the recordings represent an 'encounter between something that is quintessentially theatrical, with the cinematic medium' (Peña, 2004).

Indeed, in each instance the plays were 'respected' (in other words, mainly uncut) in a way that such texts usually are not in outright adaptations. And whatever the actual relationship between the films and an actual, original or only hypothetical stage production, almost all the AFT films are somehow very recognisably filmed theatre. This is noticeable in terms of the residues of stage acting technique existing within the performances, inevitably coming across as stagy when presented on film; the pacing of the productions, which can be extremely slow by the standards of cinema; the setting and design, which is often barely opened out from single interior settings, as if a fourth wall were added to a stage set; or simply in the emphasis and concentration on text and language. Even if not recordings of theatre performances, they are filmed almost as if they were. The results are some curious hybrids, located somewhere between theatre and film.

This appearance is in part the result of the kind of plays that the AFT chose to record – many of them linguistically complex and developed, very playwrighterly plays – and in part a result of the set of value positions that drove the project, particularly the 'respect' granted to the plays and to theatre as a prestigious form. In this, the AFT was attempting a certain kind of 'high culture' filmmaking, with the ambition to present not merely the best plays and the best performances but also the best of culture. The ambition was to do good by enabling new and wider audiences the opportunity to see these high cultural works.

Such cultural evangelism is a theme that runs through many of the attempts to conceive of some kind of film showing of stage theatre. Consequently, as with AFT, it is often the very act of representation, more than the actual content or form of any of the individual examples, which communicates most clearly the values and social and cultural judgements behind the projects. Additionally, however, as with many of the examples of theatre and dance representation explored elsewhere in this book, an underlying motivation behind the AFT project – and one that to an extent overrides the more immediate and utilitarian desire to access wider audiences – was simply that of the documentation and preservation of performances. For the AFT sought primarily to preserve records of self-consciously identified great performances, recording seminal, definitive productions, presenting plays 'how they should be done' – and in terms of the urge to save live performance from disappearance, this desire to fix a perceived 'definitive interpretation' is a notable specific example. So for Edie Landau the project offered artists 'a chance of a lifetime to record for posterity something they all felt very strongly about', while Peña points out that 'among the things film can do is preserve a great performance. And that is one of the thing I think the American Film Theatre is valuable for' (2004).

The Homecoming

A good example through which to explore the AFT's somewhat peculiar, but not necessarily unsuccessful, productions is Peter Hall's film of his own stage version of Harold Pinter's *The Homecoming* (first performed on stage in 1965, filmed in 1973). Featuring most of the original stage cast,[4] this film brought together the talent (Hall, Pinter, Ian Holm, Paul Rogers), prestige (the Royal Shakespeare Company) and high cultural values of the theatre world, precisely the kind of qualities valued by the AFT.

In interviews Peter Hall recounts his nervous reaction to being approached to make a film of his stage production, partly because in contrast to the ambitions of the AFT he describes himself as largely uninterested in preserving and recording his theatre work, which he is happy to see disappear 'like soap bubbles' (Hall, 2004). Additionally, Hall did not believe that *The Homecoming* could be made into a film, at least not in the conventional sense of opening it out cinematically. This, he argues, would dilute the power and claustrophobia produced by the play's particular theatrical setting. His conscious decision, therefore, was to confront the fact that the play was a play directly by, in Hall's words, 'endorsing the extreme theatricality of the text': 'I would regard this *Homecoming* neither as a play, nor a film. I tried to make it a play that used certain techniques of film in order to make it a play' (2004).

With this in mind, Hall's film consciously made very limited changes to the original stage production. So, for example, the stylised interior setting of the play was maintained, including employing a self-evidently muted palate and with an exaggeratedly long staircase. Very few scenes cut away from this single room interior, used almost as if it was the stage setting with the fourth wall added. Similarly the dense and verbally intense language was not loosened and re-shaped for the film, indeed none of the text was altered whatsoever. Instead, almost so as to draw attention to this theatrically ambiguous and loaded language, Hall describes how he cut shots on lines of dialogue rather than on the visual rhythm and action. Finally, the film used deep-focus cinematography to enable all elements of the screen to remain in focus at once and thereby match the sharpness and range of the human eye.[5] One result of this is that the camera does not direct the spectator's gaze in quite as absolute a manner as is the tendency with film, where focus combines with framing to effectively determine the centre of attention.

In the screen representation of live performance this selectivity of frame and determining of focus is a major point of concern, not least because every choice must at the same time exclude any alternative. As

Jonathan Miller writes 'each shot momentarily precludes the possibility of seeing anything else that is happening at the same time' (1986: 65). In contrast the theatre audience sees the action in its entirety, and while their gaze may shift from one part of the stage to another all the unattended parts remain visible and, as Miller argues, subliminally modify the experience of the salient ones. For those desiring the documentation of an original live performance event this selectivity presents a problem to be overcome.[6] However, from the perspective of analysing performance representations the choice of framing is an opportunity, revealing what was considered to be of value about the performance in the techniques and choices that were made in representing it. In terms of *The Homecoming* the choices are consciously theatrical in inspiration and impact. Here the camera still frames and directs attention in a manner distinct from the theatrical experience, but nonetheless the viewer's eye is invited and able to wander across a screen that is equally possessive of sharp and salient information. As a result, the audience is required to do much hunting and choosing of their own, dependent on the skill of the director and performer to draw and fix their attention, and thereby watching in a manner more analogous with that of the theatrical experience than of film.

Indeed, *The Homecoming* is very much rooted in a theatrical imagination; it is a play that many people would hesitate to adapt for the screen. Surprisingly, therefore, much of the success of this filmed version is the result of the deliberate attempt to maintain these distinct qualities of theatre in a different medium. The resulting experience for the viewer rests in two places, each located in oddly contradictory positions within the artefact. Firstly, it is an engrossing film, with the slightly peculiar production qualities heightening the peculiar discomfort and shock that is prompted by the action. Secondly, it is an engrossing record, particularly in terms of Ian Holm's performance as Lenny, providing the video artefact with an archival value and frisson of excitement residing in the tape's sheer existence.[7] This is something that is particularly encouraged by the history of these AFT recordings, which largely disappeared from circulation soon after their production and have only recently been made available through DVD release. As with Burton's *Hamlet*, this provides an almost legendary status to the recordings and a particular quality to the viewing experience, as Peña observes, we are 'seeing something that hasn't been available for a long time' (2004). To an extent this archival atmosphere and tone to the viewing is enhanced by the playback quality of the recordings, which sometimes includes very washed out colours, loss of detail, scratchy or thin sound and dark lighting.

Indeed, with some other films in the AFT collection, the experience of watching is not dissimilar from that of viewing the video documents of live performances described in Chapter 5.

These two qualities of viewing – as archival artefact and as performance event – do not necessarily conflict or contradict each other. Nonetheless, the way in which these recordings should and could be watched remains obscure. They cannot be watched purely as films, as the experience is continually accompanied with a remembering and excusing of the nature of the recording and of its hybrid nature. As Mark Lawson put it in a radio discussion on the AFT collection, it is difficult to imagine watching the recordings unless you already had some interest in the theatre, with a cinema-goer likely to go 'out of their mind' from the long, slow sequences of nobody speaking and nothing happening (2004). Instead, although from a strictly archival perspective compromised by the lack of a direct relationship with the stage 'original', the value of the recordings is always also archival.

Indeed, it is notable that although produced with the ambition of making the plays and performances available to wider audiences, these recordings were rarely seen at the time and in fact failed to generate large new audiences for the work among people who had not seen the original performances.[8] Re-released today, their archival value and cachet has increased with the passing of time, and no doubt they will find new if still fairly limited audiences. That the value of these recordings resides equally in their brute existence and in their experiential potential firmly places these screen reworkings within the discourse of documentation and disappearance – providing live performance with both a site of enduring presence and a marker of its enduring absence.

Video dance

In terms of their style and intent, the productions of the American Film Theatre occupy a space between theatre and film. As such these film/theatre productions are problematic because, while a strong textual discourse has been developed around the relationship between theatre and film, the AFT collection does not follow the recognised norms of screen representation. As Peter Hall puts it, *The Homecoming* breaks many of the rules of stage-to-film adaptation with the result that the recording disrupts our expectations and viewing practices and challenges our understanding of the relationship between form and medium.

In contrast to theatre, dance possesses a much more ambiguous relationship between its live and non-live representation. One illustration of this difference is that dance remains linguistically defined as dance whatever the medium of its presentation. There are no established words to differentiate live dance and non-live dance that have the same relationship as that existing between theatre and film.[9] Indeed, while rare endeavours to coins new phrases for theatre on film, such as theatrofilm or electronic theatre, strike discordant notes in their peculiarity there are numerous and competing attempts to label as distinct non-live dance. These include the obvious – video dance, dance on camera, mediadance, screendance – and the more inventive – cine-dance, choreo-cinema. Additionally, while many academic explorations of theatre on video possess a wary attitude, concerned that one medium might subsume the specificity of the other, in contrast books on video dance (such as Dodds, 2001, Jordan and Allen, 1993 and Mituma, Zimmer and Stieber, 2002) are concerned with shaping the possibilities of a new form and adopt a largely positive tone.

One result of this difference in attitude is that both the practice and interpretative field in relation to dance on screen remain in a position of flux and negotiation, consciously articulating itself as a hybrid form, in constant development and transition. Within this discourse one of the dominant themes has been the articulation of dance on video as a site of resistance, whereby the new hybrid form challenges the conventions and expectations of both live performance and screen media, particularly to the dominant screen practices of narrative, character and realism.[10] So, for example, Dave Allen writes that 'the aesthetic qualities of dance in its various forms fits less easily within the dominant storytelling practices of cinema and television' (1993: 33); while Rosenberg similarly argues that dance on screen is often 'at odds [with] "screen" as we have come to know it' (2001).

A sustained presentation of this position is also made by Sherril Dodds in *Dance on Screen* (2001), who argues that the well-established formalist and abstract practices of contemporary dance, and the instinct to present multiple perspectives and developments on the same theme, contrast with the dominance of narrative structure in film and television. Dodds also stresses the strong tradition in dance, and even more so in video dance, to be self-reflective and to focus on its own form. And, finally, in relation to postmodern dance in particular, she describes a common lack of linear or logical progression, a free play of signifiers and a focus on repeated, fragmentary moments that, when imported into video dance, mark a conscious attempt to 'draw attention to, subvert and deconstruct' dominant televisual codes (2001: 105–12). For Dodds, therefore, video dance is a hybrid site, a 'fusion or amalgamation of two distinct sites in which the codes and conventions of each medium are inextricably linked' but which also enables video dance to transgress the concepts of both dance as live performance and of screen media.[11]

In the context of this chapter, screen originating video dance works represent a step too far from the core exploration of the documentation and representation of live performance. Instead, one final example of screen adaptation examines the reworking of a live dance performance for television, exploring in particular its location as a dance work on screen and asking if the meeting of form and media does indeed deconstruct and resist dominant televisual grammar and form, or whether there remains a tendency for viewers to subsume the screen representation amongst their other televisual experiences.

DV8

Over several years British physical theatre company DV8, and their artistic director Lloyd Newson, have established a reputation for both live work and film versions of their stage performances – in particular, *Dead Dreams of Monochrome Men* (stage 1988, screen 1989), *Strange Fish* (1992, 1992), *Enter Achilles* (1995, 1995) and most recently *Cost of Living* (2003, 2004). These dance films have won numerous prizes, garnered the company much attention and prestige and also, when broadcast on terrestrial television (on both the BBC and Channel 4 in the United Kingdom), reached potentially far larger audiences than their stage performances. For DV8 the films are not only a record of their work but also a way of disseminating the productions to a wider audience, as Newson puts it, 'I am a populist. I make films because I want to push beyond the little dance ghetto audiences' (Hoggard, 2003).

In this context the status of the films in relation to the live work is ambiguous: in each case there is a stage precursor that is in some sense the original incarnation, although it would not be accurate to label it the definitive version. Indeed, in no sense are the films recordings of the stage performance, instead being 'versions', variously described by the company as 'adapted for the screen' or 'loosely adapted from the stage'.[12] Each film is clearly made *for* the screen; they are very much opened out from the theatre in terms of setting and recomposed in terms of form and technique. The films are shot on multiple, interior and exterior locations, are edited filmicly from multiple shots, use camera effects such as slow motion and dissolve, and employ familiar screen grammar, such as using close-ups to reveal characters' emotions and inner thoughts.

While reconceived on film, however, by no means does everything about these dance or physical theatre works cohere to screen expectations and norms. Examples of this can be seen in *Enter Achilles*, where the performers are constituted as a thematic group through a repeated movement motif (pulling their jackets on in an exaggerated, arms swinging motion), a structural technique that is familiar to dance but less common on screen. Similarly, although the camera does frame and direct the viewer's gaze, the movement frequently goes off frame (both left, right, up and down) as fleetingly a torso, arm or leg fills the entire screen. In doing this the screen frame is broken, not always providing viewers with the whole picture but instead self-referentially drawing attention to its composed and partial nature. In each of the films there is also a resistance to straightforward storytelling and characterisation: links are made thematically or symbolically, not necessarily sequentially or by narrative. *The Cost of Living*, for example, quickly establishes a coherent mise en scene or thematic environment (the world of seaside, pier-end attractions and performers) and develops motifs of character and relationships, but largely resists attempts to construct an overarching narrative. Instead much of the film is imagistic, constructed of short screen moments and sequences that resist reading as a sequential flow but instead form a collage-like experience.

Connected to this, there is also a willingness in the films to present the ambiguous (or be simply obscure), in a manner that reminds viewers how seldom such qualities are experienced in mainstream television. Indeed, Rob Stoneman, formerly a commissioning editor at Channel 4, suggests that along with explicit sex and explicit politics, the 'distracted culture' of television viewing cannot accept ambiguity (in Elwers, 2005: 136). In which case it is interesting to speculate whether the ambiguity of *The Cost of Living* – contained within quirky, incongruent imagery that is often presented but not maintained or elucidated – becomes more problematic in this context of being television, than when it is shown as a film at screen dance festivals, purchased and watched as video or seen in the context of its holding in performance archives and libraries.

Indeed, the particular difficulty faced with such works when broadcast on television is that they have moved from the specific, invited and actively interested audience of the live performance, archive or specialist festival, into the realm of the general and disinterested viewer of mass media. As discussed earlier, distraction can be conceived as a central fact of the televisual experience, with the disattentive viewer able to follow the broadcast flow through the recognition of deeply engrained conventions and expectations, which these films go a long way to disrupt.

Consequently, as suggested by Dodds and Allen among others, perhaps there is an almost inherent antipathy between video dance and conventional screen codes, a possibility making it worth exploring what impact this has on the experience of watching these DV8 dance films. To explore this question, discussion focuses on the most recent example, *The Cost of Living*, and its broadcast on Channel 4 television on 29 May 2005.

Television broadcast

In much writing on television culture there is an emphasis on the broadcast flow (especially Williams, 1974), a continually renewing stream of information that viewers watch 'as if live'.[13] The experience of this televisual flow represents an immediacy, a nowness, which can be equated with quasi-liveness. Sean Cubitt for example stresses how 'the discourse of TV flow is "present" in the sense that the viewer can enter into dialogue with the screen'. In other words, television proposes an absolute presence – 'here and now, for me personally' – being *for* the viewer in a direct form of address that evokes liveness (1991: 29–30). This, it is suggested, is the case even when programmes are not broadcast live, as Philip Auslander writes, 'the definition of TV as an ontologically live medium remains part of our fundamental conception of the medium – even though TV ceased long ago to be live in an ontological sense, it remains so in an ideological sense' (1999: 12).

More problematic is what impact home recording (through the VCR and more recently hard-drive PVR) has on this 'as live' experience of television. While Auslander argues that the VCR has enhanced perceptions of the relative liveness of broadcast television, for Cubitt this metaphysical assertion of presence is not only contrasted by video but also deconstructed by video:

> Liveness in a sense serves to mask the fragmentary nature of television, we might argue in reply: videotape forces back on to broadcast its own incompletion. Transparently recorded, video establishes a new relation with the audience, one that alters, I believe, the very way we watch now. (1991: 36)

Additionally, the possibilities of the hard-drive PVR, and the ability to pause, rewind and re-watch 'live' television, further erodes the cultural construction of a quasi-presence within television. With PVR, the retention and recording of the screen broadcast becomes not just a possibility but also an immediate fact, with viewers no longer needing to watch with an 'as live' level of attention and engagement. This suggests that the metaphysical perception of liveness in television broadcast can be seen as a contingent characteristic of a developing technology and in this sense very different in nature to the liveness of live performance.

Certainly the possibility of recording and timeshifting programmes abruptly interrupts the broadcast flow – we do not need to watch programmes when scheduled, as live, but instead can construct our own viewing patterns, can record, re-watch and fast-forward using the distinct cultural practices of video. I certainly did not watch *The Cost of Living* within the flow of broadcast television, but recorded it and watched (and re-watched) it several days later. The scheduling of dance and performance films, which are typically targeted towards minority audiences and shown in late night slots, particularly encourages such habits.

To an extent my experience of *The Cost of Living* was located within video culture and therefore removed from the televisual flow. However, the experience was not entirely disconnected from that of television, not least because, with present technology, home recordings continue to be interrupted and framed by adverts, channel indents and programme trailers. Even if the broadcast flow is subverted, an echo of its disrupted presence exists in the still present trailers for programmes that are now in the past. We will, additionally, continue to watch the film through our cultural associations of the station identity (here Channel 4), and with whatever particular expectations and perception of value associated with such a marginal work having made it onto mainstream television. This might be negative – with thoughts of self-censorship, or the perception that 'Video for TV has to be more palatable and more cosmetic. You're trying to reach a mass audience' (Lynn Taylor Corbett in Billman, 2002: 17) – or positive, with the perception that such broadcast confers a significant marker of quality.

Hybrid spectatorship

While aspects of the experience of watching *The Cost of Living* resist screen grammars and conventions, at the same time the film also contains and utilises several markers of a televisual, 'as live', metaphysics of presence and fabricated intimacy with the viewer. Examples of this include the direct address to the audience at home and the use of the performers' own, real names. In one scene a performer (David) ends a dialogue made directly to the camera by moving off-screen while saying 'I'll be right back, don't go away'. The broadcast immediate cuts to an ad break. The convention alluded to here is that of the chat show host or newsreader, with the pretence of entering into our living rooms in an artificially constructed echo of a reciprocal relationship ('Don't go away, we'll be right back'). Partially ironic homage, in this context this device has been shifted from its familiar use in news or light entertainment programmes (which often *are* broadcast live) into the more incongruous setting of a pre-recorded drama. However, combined with the nature of the material within *The Cost of Living* – not least that the performer/character David has no legs, with our responses to this foregrounded by the film's explicit interrogation of his physicality through dialogue and implicit examination through movement – the result is an assertion of a kind of metaphysical presence, a thereness-for-me of the performers that emerges from the closeness provided by the camera.

This use of conscious televisual techniques to assert a kind of quasi-physical presence to the film experience combines with another factor fairly common to video dance that also seeks to construct an immediate relationship with the viewer. This is the presentation of dance in real world locations (in *The Cost of Living* these range from the pier, to a row of lock-up garages, with the film described by one critic (Parry, 2005) as filmed 'on location in the real world') rather than in the familiar back-drop of theatres or rehearsal rooms. The effectiveness of this device is based upon the incongruity of seeing stylised and choreographed movement in the context of an everyday location. It plays upon the perceived mismatch between the grittiness of the real world and the abstractedness of dance, challenging our attempts to make sense of the film according to expectations of realism (evoked by both the locations and the medium itself), calling instead on engagement through metaphor and the theatrical imagination.

The Cost of Living also contains much else drawn from the techniques and syntax of television and film. The film opens, for example, with the camera panning over the beach and pier location, employing the familiar television grammar of scene setting. Structurally familiar, suited to viewers' media-derived expectations and attention – and therefore also fairly short, swiftly getting to dialogue and the central characters – this compares dramatically with the often painfully slow beginnings of the AFT films examined earlier and video documents of performance explored in Chapter 5. Finally, it is worth noting the technical quality of DV8's dance films, their high production values, their visual and editorial sharpness, the picture quality and the full, bright colour reproduction. Unlike the other examples explored in these chapters they can be watched for their own sake, rather than needing to be interrogated, excavated and made excuses for.

Yet at the same time watching *The Cost of Living* is also characterised by distraction. Despites its copious use of the techniques and tools of television, the impossibility of reading the film solely within familiar screen grammars means that it is demonstrably harder work than watching most screen media. This is not purely the result of content – of the difficulty of the subject – but also of the combination of form and medium, of the simultaneous use of and resistance to televisual codes and conventions. In other words, if DV8's video dance works do occupy a kind of hybrid space, located between live dance and conventional screen practices, then so are viewers' watching experiences located in a divided or compromised position.

As viewers, our relationship with these films is problematic. It is clearly inappropriate and impossible to attempt to see through the screen, using it as a transparent document to a live work located elsewhere. Yet at the same time the codes and conventions required to read the films are not entirely located within the screen media itself, but also reference and echo past manifestations and a theatrical, live imagination. The problem, yet again, is a difficulty in knowing how to watch these screen representations of performance.

Watching screen representations

Both this and the previous chapter have sought to think around the subject, or question, of live performance on video. Together these two chapters have explored possible aesthetic sensibilities, narratives of production and cultural practices of use that characterise video as a medium. More crucially, the discussion has explored video as a culture, as an activity, as a verb and as a way and type of doing – thinking about what it means 'to video', both in terms of to record and to watch. In drawing this exploration to a close it is worth focusing on the watching strategies that are employed for and produced by the experience of seeing and knowing live performance through its screen representations.

Interrogating my own experience of watching various representations of theatre and dance performance on screen, across a range of relationships (from video document to screen reworking), I recognise a distinct ambiguity and uncertainty in my attitude and practice. Notably, this is a perplexity that is rarely replicated when looking at photographs or reading reviews of performances, the subject of the following chapters. With live performance on screen, in contrast, I frequently find myself questioning how I am supposed to be watching: whether I am allowed to notice the technical deficiencies (or whether I am obligated to make sense of them, to read them into the mise en scene), whether I can view outside my narrative-based, realism-dominated televisual literacy, whether I can watch with the heightened level of attention encouraged by live performance as opposed to the disattention of video culture, whether use of the fast-forward button is laziness, or part of the full utilisation of the medium. I also find myself confused about what kind of experience I should be having, about whether it should be enjoyable and about whether I can see through the transparency of video to the absent live performance.

As has been explored, a significant part of this difficulty descends from the fact that screen media are (like live performance itself) modes of performance presentation in their own right, and moreover ones accompanied by a rich and powerful set of established cultural practices and aesthetic modes. Additionally, these modes of making and watching screen performance often conflict with those of live performance. Yet at the same time one of the particular modes and practices of video is of transparency, and the assumption (unconscious, naïve, unthinking but nonetheless significant, seductive and enduring) that the camera records neutrally, passively and absolutely. Together these two factors problematise our viewing experiences.

To an extent our attitude should be determined by each different relationship that exists between individual screen representations and any original or exterior live performance. Commentators such a McAuley and Melzer describe the need to develop a new kind of literacy and interpretative strategy for reading diverse forms of representation. We should watch video documents such as those of Forced Entertainment differently from screen reworkings made for broadcast (such as those by DV8) and from adaptations made for wider public screening (such as the American Film Theatre productions). This is self-evidently right. Nonetheless, in each instance I find it necessary to view these screen artefacts according to the extent to which they either resist becoming video, become subsumed within televisual culture or alternatively actively seek to stretch and alter our perceptions and readings of screen media.

Marcelle Imhauser suggests of the screen representation of theatre that 'The most perverse impact of television is to turn everything into television' (1988: 97). In many ways this is true, as we watch the screen directed by televisual viewing habits, with our judgements at least partially informed by our screen literacy. Sherril Dodds, on the other hand, suggests that video dance 'calls into question our expectations of the televisual medium' (2001: 120). I also recognise that this can be true, and not just for the postmodern video dance Dodds is specifically referring to but also to the way in which screen representations of live performance frequently resist being viewed as or becoming television, not least in the sense that they frequently fail to become *good* television. Indeed, a perverse usefulnesses of the distortions resulting from the failure of the video documents to make good (televisual, watchable, entertaining) recordings of live performances is that they do resist turning the performance into television. It would be almost comically impossible to imagine many video documents of live performance, such as those of Forced Entertainment, being broadcast on television. The difficulty is that both these aspects and relationships seem to be true at one and the same time, making it necessary to think a little longer about what meanings and perceptions these video representations communicate, not so much about the individual performances represented, but instead about broader understandings of the relationship between live performance and its screen representation or replication.

Within postmodern cultural commentaries various writers, including Philip Auslander but also Steve Connor (1989), Roger Copeland (1990) and Steven Wurtzler (1992), have taken Jean Baudrillard's exploration of the relationship between the real, the copy and the hyperreal and applied its terms and discussion to the relationship between the live and the non-live. Auslander, for example, suggests that 'In a special case of Baudrillard's well-known dictum that "the very definition of the read is *that of which it is possible to give an equivalent reproduction*" (Baudrillard, 1983: 146), the "live" can only be defined as "that which can be recorded"' (1999: 51). Yet watching the screen live representation of live performance seems to highlight the nature of the difference, the disjunction between form and medium, the gap or absence where the live once resided. Indeed, it seems tempting to suggest that the live is located within this gap, defined precisely as that which cannot be recorded.

In which case Auslander's other key argument, that the very attraction and cultural valuation of live performance is something created by mediatization (1999: 55) remains central and valid – we can only feel that something is missing as a result of watching the recording. As Wurtzler also argues, the demand for and valuation of live performance is something in part created by a culture in which the live has been encroached upon by the non-live:

> Rather than the 'death of the aura' at the hands of mechanical or electronic reproducibility, the recorded reinstates 'aura' in commodity form accessible only within those events socially constructed as fully live. (1992: 89)[14]

Connor echoes these arguments in *Postmodernist Culture*, describing live performance as a socially produced object of desire. Indeed, for Connor the proliferation of reproductions and media representations increases our cultural desire for origins: the more something is replayed, the more it affirms that it is not the real thing, the more we want the real thing. 'In the case of the "live" performance,' Connor continues, 'the desire for originality is a secondary effect of various forms of reproduction' (1989: 151–3).

This perceived value of the live and desirability of liveness in contemporary western culture is frequently in evidence, witnessed by the repeated use of the talismanic word 'live' in advertising and branding: live football, live in concert, recorded live, coming to you live from, in front of a live studio audience and so on. Such usage constructs a number of equations, with different emphasis on different occasions, between the liveness of the event and its genuineness, its realness, vibrancy, urgency, uniqueness and, ultimately, value. While the greater an event's claim to be fully live, the greater its ability to claim the attendant ideas of 'aura' and desirability, these claims are made even when the event is not absolutely live. Liveness itself here becomes a commodified concept, something that can be packaged, bought and experienced, much like any other commercial exchange.

While made visible and valuable by mediatization, however, the manner and struggle inherent in the attempt to represent live performance on screen media reminds us that the live is a particular medium of its own – it is its own kind of technology. As a result of either deliberate intention (as with video dance), through the side effect of technological failure (as with video documents) or the evident mismatching of form and medium (with the American Film Theatre productions), the act of watching and re-presenting live performance on screen generally, if not always, disrupts normal screen practices. This disruption is useful in reminding us of the constructed nature of those practices in the first place, also making us conscious of the different viewing strategies required to watch live and non-live performances. Screen representations of theatre and dance performance, therefore, make an implicit statement about not so much the loss or absence of liveness, but about the otherness of liveness. They assert that the live is a medium, a technology and a verb in its own right.

Part III

7
Photography, Truth and Revelation

For the last century and more, still photography has provided the most commonly used medium through which to know and see theatre and dance outside live performance. The flexibility, economy and unsurpassed reproducibility of photography means that this continues to be the case today, more visible and widely reproduced than any other medium of performance representation.

As with other media of representation, the relationship between photography and live performance is articulated through the recurring discourse of disappearance and documentation: the photograph records performance for posterity, but also fundamentally transforms it into a different artefact. It is the tension existing in this dual role that forms a central strand of this exploration of performance photography, investigating how contemporary photographers, working in collaboration with directors, choreographers and artists, have employed the medium to represent the movement and narrative, the temporal and spatial existence and the energy and liveness of live performance. To root the discussion in practice, the analysis is focused and exemplified through the consideration of a range of theatre and dance photographers, whose work demonstrates a range of contrasting techniques and motivations, and different practical and aesthetic relationships, with theatre and dance performances.

The two chapters in this section, therefore, look at the ideas and knowledge that different photographers communicate about both the particular performances they represent and also about live performance more generally. First, this chapter explores conceptualisations and practical illustrations of performance photography as a revelatory (although not always documentary) medium, a perception fundamentally calling on our faith that the camera shows us 'what has been'. This chapter draws examples from the work of four photographers, namely: Graham

Brandon (archive photographer for the Theatre Museum, London), Dona Ann McAdams (New York photographer of performances at PS122), Bertien van Manen (Dutch photojournalist) and Lois Greenfield (American dance photographer and 'photo-choreographer'). This discussion explores how photography is at once seized upon as a medium that will document, and thereby save ephemeral theatre and dance performances from disappearance, and at the same time is lamented for its inability to record the complete appearance or capture the actual experience of the live performance. Instead, performance photography becomes a transformative art form in its own right that seeks to reveal more than the surface appearances of performance.

Continuing this perspective, the second chapter in this section considers photography as a more overtly representational and transformative medium, particularly in the context of its use as a promotional and marketing tool, considering the working of four further photographers, here: Chris Van der Burght (Flemish theatre and dance photographer), Hugo Glendinning (British dance and theatre photographer and close collaborator with Forced Entertainment), Euan Myles (Scottish-based advertising photographer) and Chris Nash (London-based dance photographer). This chapter argues that the centrality of selection to the act of photography – a centrality that reveals the photograph to be fundamentally transformative, constructed and representational – should be central to the analysis of all photographs. The still image must always enact a distinct interpretation, a selective construction, which in its choices, omissions and creativity tells us more about attitudes to and understandings of performance than merely pointing us towards what it purports to show.

Alongside the discussion of the work of specific photographers, the following two chapters also consider salient points relating to photographic theory and practice – particularly in terms of the economics of production and the narrative of use that direct much performance photography. This chapter here begins, however, by exploring what is undoubtedly the most enduring cultural conceptualisation of photography – as an authenticating and revelatory medium. Following sections then draw out the significance of this revelatory ontology in terms of the use, conceptualisation and practice of performance photography.[1]

Aesthetics of revelation

As a medium that can create permanent records of transitory events, the history of theatre and dance shows how quickly photography was seized upon for purposes of documentation. As Laurence Senelick comments on photographs of performers from the 1860s and 1970s, 'what more natural than that the actor should register a salient moment from a performance, in full costume and make-up?' (1987: 5). This documentary motivation is further crystallised in a remark by Anna Pavlova in the 1910s, who declared to a photographer 'my art will die with me. Yours will live on when you are gone' (in Ewing, 1987: 14). Artists, scholars and historians, conscious of the ephemerality of theatre and dance, often seek a method of representation that will still the transient and capture the complete appearance of live performance. With greater claims to authenticity than illustrations or non-mechanical representations, still photography appears to offer this possibility, calling on the existence of a widespread cultural understanding of the photograph as producing an image true to the appearance of the real world and revelatory of something that existed. And indeed, today such nineteenth- and early twentieth-century photographs of actors, dancers and singers do provide us with an alluring glimpse of past performers and performances that might otherwise have vanished.

This faith in the intrinsic relationship of the photograph with the real was born with the emergence of the technology in the early nineteenth century; texts on the history of the medium invariably overflow with testimonies to the documentary pre-eminence of the photograph. Witnesses include pioneers in the field, such as Daguerre, who asserted that photography was a 'process which gives [nature] the power to reproduce herself' (in Marien, 1997: 3), along with Lumière, who declared that he 'only wished to reproduce life' (in McQuire, 1988: 15). And during the course of the twentieth century a series of hugely influential commentators have continued to assert the primary revelatory aesthetic of photography. Central among these is André Bazin, who affirms an 'irrational power of the photograph to bear away our faith' (1967: 14), along with Susan Sontag, who writes in *On Photography* that 'The assumption underlying all uses of photography [is] that each photograph is a piece of the world' (1979: 93), and Roland Barthes, who declares in *Camera Lucida* that the photograph's primal force lies in its unsurpassed power of authentication and ability to declare 'that has been' (1984: 81). Today there remains a tendency to regard the photograph as revelatory and authoritative: we prove our existence with passport photographs, record our memories with snapshots and trust such images to capture the world as it really is.

There is, therefore, a great weight of cultural convention and conviction that prompts this strong instinct to respond to photographs on at least a quasi-documentary level. Yet it is not as straightforward as that, with qualifications as to the authenticity of the photographic image significant and well-known, including inherent limitations such as the restriction of the image to two-dimensions and the distorting and flattening effect of the camera lens. Photography also transforms its subject through the effects of lighting, shutter speed, camera angle, colour distortion, methods of print development and the results of cropping or shot selection. And processes of photographic selection or transformation are joined by more interventionist possibilities of manipulation, editing and fakery. Through elements both inherent to the medium or caused by the circumstances of its employment it is evident that the camera does not directly reproduce the world but instead transforms it into photography.

Those eulogising photography for its power to reveal the world, however, do so not because it actually reproduces the world, but rather as a mark of its dominant position in our culture. Barbara Savedoff, for example, recognises the limitations of photographic authenticity, declaring that despite an 'aura of objective accuracy' a photographic reproduction always distorts what it presents. Nonetheless, whilst questioning the documentary power of the camera in fact, Savedoff also reaffirms the importance of our perception of photographic authority: 'Whether it is warranted or not, we tend to see photographs as objective records of the world, and this tendency has far-reaching influence on interpretation and evaluation' (2000: 49). Unless directly prompted towards doubt, by evident fakery in appearance or impossibility in content, we faithfully continue to equate the photograph to the real.[2] Indeed, for Savedoff, the 'perceived special connection to reality can account for [photography's] distinctive aesthetic impact' (2000: 8) – in other words, the essential aesthetic quality of the photograph is that of revelation.

The central implication of this revelatory aesthetics is that the tendency is to respond to photographs according to how they present the world. As Sontag writes, 'what a photograph is *of* is always of primary importance', with it being impossible to consider a photograph without considering its subject in the world, without considering what 'piece of the world' it is of (1979: 93). In the particular context of performance photography discussion will later explore precisely how problematic this preposition 'of' becomes. Before that, however, the first detailed example of performance photography in practice looks at the work of a photographer operating in a world inherently defined by documentation, truth and a revelatory aesthetics. Yet even here the constructed and multiple nature of the photographic image is revealed as central and inherent.

Archive photography

The perception of the camera as authoritative recorder of performance is made concrete in the concepts (if not necessarily or entirely the practice) behind the work of Graham Brandon, archive photographer for the Theatre Museum at the Victoria and Albert Museum in London. Alongside other materials, including notably video, the Theatre Museum collects and commissions still photographs as part of its wider archive of theatre and dance in the United Kingdom. In doing this the archive seeks to preserve ephemeral performance and provide as complete and transparent as possible access to performances of the past. As explored in relation to archives in Chapter 2, there are inherent challenges to such archival endeavour: not least finite resources and practical limitations that mean that not all performances can be recorded. Here, however, the focus is on what happens after any initial selection and collecting decisions have been made, and on what sort of photographic records of performance are anticipated and constructed.

On this, Brandon articulates a series of approaches and techniques that are prompted by the documentary ambitions of the Theatre Museum and which direct his photographic practice. In particular he speaks of the need for the archival photograph to pull back from the performance and produce an image 'with space around it', showing the performers in the context of the set and stage.[3] In many ways this approach is the still photographic equivalent of the video recording of a performance taken by a static, single camera located at the back of the auditorium: each producing largely opaque and distant records that are dissatisfying in their own right, but which we can learn to interpret to elicit factual (if not emotional) information about the performance. Further, just as video documents of performance frequently conflict with established televisual grammars and expectations, so do these archival images often contradict the demands and aesthetics of the photographic medium. As Brandon himself admits, these images rarely contain much in the way of photographic power as images.

In live performance the audience's relationship with the stage is dynamic, flexibly constructed by the choice of what to look at and concentrate on at any particular moment. Composed to capture the whole stage image, archive photographs seem to mimic the actual physical appearance of the theatre – with the frame of the image replicating that of the proscenium arch of the stage. Aside from the resulting smallness of the image, however, this also elides our experience of photography, which in contrast to the flexible gaze of live performance is through the whole composition of a fix-framed image. The act of photography, of constructing a framed image, must always involve an act of selection; the difficulty with these photographs is that they fail to enact this selection, instead attempting to adhere to the aesthetics and frame of another medium.

Tellingly, Brandon himself is very conscious of how the medium of photography mitigates against many of the overt archival demands for neutrality and completeness. Indeed, perhaps because he is faced daily with the practice of photography, Brandon never sets out with such clear-cut conceptual, archival ambitions in mind. For him it is unequivocal that the concept of archival photography is 'trying to do the impossible' in trying to capture performance. Instead he talks about his photographs as 'notes' from the performance: 'You're getting a smell of it, a feeling that will hopefully give an idea, a concept, of what it was all about'. The question remains as to what kind of notes of the performance one aims to take – or indeed, can take.

For the centrality of selection to still photography runs counter to the explicit objectives of archival completeness. Something that Brandon recognises in his own work, as he describes a balance between aiming to get good photographs and aiming to get good archival records:

> We try and get some long shots, to help balance ... Because it's too easy, particularly with photo-calls of theatre, you start to get closer and closer and before you know it you're just shooting portraits. And that isn't really what ... well it's not capturing the theatre is it?

At the same time, however, Brandon confesses that it is the tighter photographs that appeal to him personally: and these images do indeed make 'better' photographs, not least because they adhere to the aesthetics of photography as a medium in its own right. One example of this is a photograph of Complicite's production of *The Chairs* (1997), where the image is tightly composed around the two performers thereby isolating them from any other action beyond the frame (Figure 7.1). Unlike the widely framed archival photographs, this image is more selective and overtly interpretative, directing viewers towards a particular kind of knowledge of the performance. In this image an intense relationship between the couple is suggested by the close cropping of the frame, with the similarities in their age and their facial expressions uniting them in a conspiratorial bond. Viewers cannot know if this is a 'true' reading of the production, yet at the same time it seems to communicate an emotional truth and a powerful theatricality. It allows us to imagine and almost experience the performance, while the archival images only seem to empirically show us remote, surface appearances.

Figure 7.1 *The Chairs* (Richard Briers and Geraldine McEwan).
Complicite and Royal Court Theatre (1997). Photograph Graham Brandon.
V&A Images/Theatre Museum.

There is an irony in this, as the tightly framed photograph is entirely unlike the audience's experience of the live performance. Rarely are audiences this 'close' to the performers, and never this close in framed isolation. However, while this is not the actual perspective of an audience it is perhaps the *imagined* perspective, the feeling of emotional and physical closeness to a startling stage performance. The power of photography is in bringing any performance this close.

Indeed, in an interesting way this photographically inspired perception of closeness has the ability to impact on how we see, or rather think we see, the live performance itself. We frequently watch a performance having already 'seen' it through photographs used in advertising material. In my own experience I have noticed how tiny fragmentary moments of the performance consequently strike me with a powerful sense of the familiar; they appear disproportionately prominent precisely because the camera has depicted them. Yet I have not seen these moments before and in truth am not seeing them at all now, for the moments represented are of a photographic closeness and stillness that is impossible to see live. The moments would be entirely invisible in life, entirely subsumed into the flow of the performance, if not prompted into existence by the photograph. Yet although the images presented in the photographs do not exist (as such) in the performance, these moments dominate and define my memory of the performance.

A similar relationship exists between still photography and video, television, or cinema. As Sontag writes, any moment presented motionless in a photograph holds a discernable authority over a flow of images, 'Television is a stream of underselected images, each of which cancels its predecessor. Each still photograph is a privileged moment, turning into a slim object that one can keep and look at again' (1979: 18). Further, the selected performance photograph is in contrast to the 'underselected' flow of the performance itself. This potentially reverses any perceptions of a hierarchical relationship of original and copy, between the photograph and the event. For the viewer, the performance almost becomes an imperfect repetition of the moment presented so immediately and close in the photograph. Something similar happens when we look at photographs after watching a performance, with the images impacting on our memory and enduring understanding of the performance.

Such perception overturns the subordination of the representation to the original; it also highlights the continuing power of the still photographic image. Faith in photography to reveal the world includes or creates a sense that what is revealed is what is significant. Additionally, still representations hold hierarchical status over moving subjects, with the act of selection privileging a single moment against the unselected whole. Despite, therefore, the greater claims of completeness of other forms of documentation (particularly video), the representation of performance by still photography continues to constitute the greater part of our imaginings about performance.

Instead, therefore, of possessing an archival authority, much performance photography has a different kind of authority, which is of intimacy and getting far closer to the performance and performers than an audience ever is in practice. Here the aesthetics of *revelation* is truly enacted: the camera showing us the world anew, a world that is not passively found but actively constructed, framed and transformed by the act of photographic representation. This photographic 'revelation' is importantly distinct from 'documentation': the cropped and composed image not merely documenting but instead actively constructing and revealing the world. While Brandon's pulled-back archival images seek documentary passivity, his composed and interpretative images in contrast are consciously active and revelatory.

As archive images, however, the problem is that these tighter photographs are more selective and less complete in their recording of the original stage appearance. Here viewers cannot know what is going on outside of the frame; cannot access the context of the emotion communicated; cannot even pretend that they are seeing what the audience would have seen. Yet at the same time as telling us less about the performance as theatre history, these photographs do also seem to communicate more to us – and especially more to us about the performance *as* performance. The difficulty, then, is that the two kinds of image – the widely framed archive image and the consciously composed photograph – are able to claim and communicate different kinds of completeness and meaning: one seeking to communicate an emotional truth; the other a visual, historical truth. These two levels of reality frequently overwrite and efface each other, particularly as in the next example, where the factual history of the performances is intimately connected with an almost mythological emotional history and reception.

Photography as witness

The instinct and desire to use photography to halt the disappearance of performance is very strong. This is especially so with performances perceived to be particularly susceptible to disappearance as a result of their marginalised position in our culture, where the photographic image becomes verification of the existence, appearance and importance of its subject. Photographs offer an alternative and mechanical memory of performance; a validating proof that the performance, now gone, actually happened. When Rodrigues Villeneuve asks himself 'what do we expect from theatre photography?' he replies:

> I would say, naïvely, the saving of the performance, which disappears as fast as it is produced ... Isn't the photograph a physico-chemical trace of what happened at one moment on the stage? Are we not in the presence of an imprint of the theatrical real? (1990: 32)

Nicole Leclercq affirms this perspective when he describes photography as holding a 'place of faithful witness and privileged memory of the theatrical phenomenon' (2001). This role of the camera as witness can be found in the work of Dona Ann McAdams and her photographs of off-Broadway productions in New York since 1983 – and in particular her work as house photographer for PS122, for whom she has photographed every performance from 1984 onwards.

Operating in a largely un-scripted, non-object based and oral tradition of immediate and intimate live culture, performance art almost by definition only leaves memories behind. With such work often positioned as a counterbalance to a commodity-based culture, this transience can also become a positive attribute and value of active resistance to the reproducibility that documentation entails. Consequently, as discussed in previous chapters, those working in performance art become at once fearful of disappearance and at the same time wary of the transformative nature of documentation and the determining power of representation and of misrepresentation. Against this background, McAdams' work forms a photographic record of New York performance art of the mid 1980s and 1990s, constructed through the development of relationships of trust and close collaboration with the artists themselves.

What is particularly striking about McAdams' photographs is that they are taken (for the most part) during actual live performances or public 'performance views'. This is in contrast to the vast majority of performance photography elsewhere, which is largely taken during photo-calls, as is the case with most of Brandon's images,[4] during rehearsals or in the studio. With McAdams these circumstances of production result in black and white images, taken without flash and caught during the live performance, manifested photographically in grainy contrasts of light and dark. Frequently the frame is narrowly focused on the performance artist themselves, often performing solo in pieces where the individual artist is both the creator of the work and the work itself, which is inseparable from the artist's personal history, body and physical presence. These images are in many ways portraits, seeking to engage the viewer with the presence, personality and force of the performance through the intimate portrayal of the performer. For example, a photograph of Penny Arcade performing *While You Were Out* (1985) represents the performer intimately, as unabashedly, powerfully present for the viewer (Figure 7.2).

The power of McAdams' images is certainly in part the result of our cultural trust in the documentary power of photography to preserve and record the world as it is. Indeed, the images evoke a visual mode akin to photojournalism, where there is a clear invitation to trust and accept the photographic image. As Colin Jacobson puts it, as reportage photography has the ability 'to cut through the complex, fluid, contradictory nature of the world and strike a moment of truth' (Jacobson and Haworth-Booth, 1994: 32). Here truth is located in the content of the image, in what they show us of the world. And like photojournalism, McAdams' photographs seek to efface themselves, to be invisible or transparent as we look through the image to the performance itself.

Yet the photographs are not neutral or transparent. As with all photography, as with photojournalism, we should always look beyond the presentational content of an image and think about where the photographer is in relation to the picture and of the narratives of production behind the photographs (Alvarado, 2001: 151–5). Here the photographs, taken in closely collaborative relationships with the artists, are constructed and selected to display a particular view of their subject and of their world. If the aesthetic of McAdams' work possesses a rawness, boldness, directness, an assertion of presence, then this is partly the result of the circumstances of production – of shooting during live performances – but accident becomes virtue as the aesthetic of the photographs matches that of the performances themselves – which also possess a rawness, boldness and assertion of presence. In this McAdams has found an approach that allows her to record the performances in a manner apposite to their own values, form, message and style of articulation. This is a representation of social, physical, political and emotional presence, embodiment and engagement, and it is this truth – rather than a neutral or found truth – that the photographs attempt to capture and communicate.

An example of this is McAdams' photograph of Karen Finley performing *I'm An Ass Man* (1987). In this image there is something perfectly, simply satisfying about the way the microphone bisects the photographic vertical, with Finley almost symmetrical behind it – one leg on either side, clenched fist on either side (Figure 7.3). The attraction of this balance serves to emphasise the break in the symmetry, as Finley's left breast

Figure 7.2 Penny Arcade. *While You Were Out.*
Performance view at PS122 (1985). Photograph Dona Ann McAdams.

Figure 7.3 Karen Finley. *I'm An Ass Man* (Club Show).
Performance view at The Cat Club (1987). Photograph Dona Ann McAdams.

protrudes, awkwardly, uncomfortable, unapologetically, over the top of the frilly neckline of her dress. The self-effacing nature of this photograph – presenting not itself but Finley – is the result of this constructedness, not the result of neutrality. The appearance of the photograph consciously matches the ideology and aesthetics of Finley's own self-representation in performance.

Writing of photographic records of live art performances, RoseLee Goldberg positions images such as McAdams' as traces and triggers to memory and the imagination of the theatrical presence, stating that 'each picture carries the residue of time, and each has a way of making the past present and fleetingly real' (1998: 35). Similarly, Carr writes that McAdams' photographs remind her 'of that old idea that photography is magic that feeds on someone's power' (in McAdams, 1996: 8). Performance art is embedded in concepts of presence, with the artist and performer and individual appearing as one before the audience – the performances are about the staging of personality, of identity and sexuality, about the staging of the body and of presence. In live performance the experience of the presence of the performer makes a call or demand on the spectator, evoking a direct moral relationship that constructs audiences to such live art events as witness (see for example Heathfield, 2000: 105). For Goldberg and Carr photography is similarly invested with concepts of presence, with the camera now called upon as witness to the event, a direct and quasi-moral relationship that also seeks to implicate and engage the viewer of the photographic representation through the lens of the camera.

This can also be seen as the aesthetic of McAdams' images, which become acts of formal witness to the event beyond the mechanical operation of photography and the objective recording of the world. As acts of bearing witness, the photographs seek to contract us – the viewer of the photograph, not or rarely also the audience to the performance – to the photograph in the same way that the live audience was contracted to and implicated in the live performance. As witness to the live art event one becomes more than an audience or bystander; as witness to a photograph we may similarly become more than viewers.

This is what is necessary for the photographs to remain perceived as representations of the performance rather than becoming self-contained images in their own right, decontextualised and misrepresentable. This is what is necessary for the photographs not to become voyeuristic, and not, as happened with McAdams' images of Finley, to become perceived as transparencies onto something decadent, obscene and indulgent. It is here that the representational relationship between McAdams and Finley, extremely close in realisation, presents a revealing example of the potentially devastating power that the enduring photographic representation holds over the transient live performance.

In the mid 1990s the work of Karen Finley – and also other performance artists, particularly Holly Hughes, John Fleck and Tim Miller, all of whose work was photographed by McAdams – became the focus of attacks from right-wing groups in the United States who condemned it as obscene. In particular, attention was focused around how the work received support from the state financed National Endowment for the Arts. The argument ran to the Supreme Court on the question of whether or not such money should be awarded subject to consideration of 'general standards of decency'. Looking back at this case what is relevant here is how McAdams' photographs became central evidence, even prosecutory *witnesses* (now in the legal sense of the word), in the campaign of censorship and moral obduration against the New York performance art scene.

For once in widespread circulation, performance photographs are no longer under the control of the artist (or indeed always the photographer) in terms of how and where they are reproduced, nor in terms of how they are framed and contextualised. Instead, the photographs have to stand alone, with the performances perceived according to their photographic representation alone – or even within the context of deliberately censorious framing. It is here that the untethered reproducibility of the photographic representation becomes an issue.[5] For opponents, the photographic images were authoritative and accurate depictions of the performances, which could be used as legitimate weapons of criticism and abjuration. For the defenders of the work, such use and de-contextualisation was dishonest, instead constructing the images as limited and flawed traces of the live experience. This differing perception is clear in numerous accounts of the conflict, where a repeated refrain was that her detractors had never seen Finley perform (for example Carr, 2000), a complaint that articulates a moral and evidential difference between seeing the performance and seeing the photographs.

The central question here is the extent to which photographs are a legitimate tool by which to *see* performances. With McAdams' photographs today becoming historical records – as opposed to news – this use of the images to allow people to see the performances while never actually having seen the performances is precisely that to which they are put in research, in universities and in performance archives. It is precisely this use that is evoked by Goldberg, Carr and many others when they extol the historical importance of McAdams' photographs. The problem remains then as to whether viewers are able to use these photographs in a way that is not only sympathetic to the original act of performance – and therefore liable not to misuse the images for political purposes – but also that actively binds them to the photograph in a way that begins to replicate at least some of the original political and emotional significance of the performances.

What certainly does happen is that in freezing and isolating a moment of the performance the photographs present the work to us in a new and changed manner from that experienced by the live audience. The relationship is less defined, less mutual. The gaze between performer and audience is no longer returned and it is difficult, if not impossible, for viewers to retain commitment to the performance – the eye can look away or the page be turned too easily for that – and a degree of distance and detachment is inevitable. Indeed, looking at these photographs now transforms the live aesthetic of presence into an aesthetic of absence inherent to all historical records and documents. This distance, however, is replaced by a new kind of closeness, the particularly photographic closeness – very different from video, for example, and not only the result of the literal close-up – that results only from stillness.

This is the case, for example, in McAdams' photographs of Annie Sprinkle performing *Post Porn Modernist* (1990), precisely the kind of work and kind of photograph that became subject to de-contextualisation and condemnation in the mid-1990s court cases (Figure 7.4). Inevitably, such photographs are always going to be viewed first in the light of their content, with the resulting potential to implicate viewers as voyeur or user of the photographs as pornography. This, at least, was the contention of right-wing opponents to such work.

Indeed, in terms of content, this image contains many of the indicators of pornography, as does Sprinkle's work in performance, which seeks to subvert such indicators and reclaim ownership of the sexuality of her body. Similarly, this photograph transposes the pornographic gaze first by the aesthetics of the image – the black and white, grainy texture, the light spots, the overly tight framing, the caught aesthetic – which are not easily read as the visual mode of pornography and so partially circumvents this perspective. Yet more than anything else it is the ability of the eye to linger, to take in detail, which really changes our relationship with this image. All still photography theoretically allows viewers to look at an image for as short or long a time as they want, but this potential is different from the actual desire or need to look for as long as is necessary, or as is possible. Resisting an even quasi-pornographic gaze, this image draws viewers deeper, past the surface vision that is the aesthetic of pornography, to see the detail in stillness. This is only a potential,

Figure 7.4 Annie Sprinkle. *Post Porn Modernist*.
Performance view at The Kitchen (1990). Photograph Dona Ann McAdams.

depending very much on each viewer's willingness to engage with and *work* with the photograph.

Such performance work as represented in McAdams' photographs is inevitably always going to be experienced live by small audiences. This is partly a matter of choice on the part of the performers, partly a matter of its content and also a question of economics and practical restriction. Photographs of such work, in contrast, have the potential to reach far larger audiences, outside circles of sympathy and intimacy. Here they will always be de-contextualised and potentially subject to deliberately condemnatory framing. Here the willingness of the viewer to actively work with and engage in the image is always going to be more doubtful. This is part of the nature of photography, where its ability to reveal a piece of the world to us is subject to the questions of whether and how we want to see that revelation.

Theatre and photojournalism

McAdams' photographs evoke the power of photojournalism by dint of her being *there*, of being a privileged witness to transitory, now historical and often seminal performances. This authority of attendance, of present witness, is revealed again in the work of Dutch photo-documenter Bertien van Manen. However, a revealing tension and the possibility of doubt within this authority emerges through the contrasting experience of van Manen's photojournalism with her one-off involvement taking pictures not of life but of theatre.

In 1998 van Manen was approached by Alize Zandwijk, director of the Ro Theatre in Rotterdam, to take photographs of a production of Maxim Gorky's play *The Lower Depths* (produced as *Nachtasiel* in the Netherlands) (Figure 7.5). Written in 1902, Gorky's text is set amongst the inhabitants of a Moscow doss house, and Zandwijk's inspiration for producing the play came in part from having seen van Manen's photographic account of the lives of poor Russians in the 1990s – published and exhibited under the title *A Hundred Summers, A Hundred Winters* (1994) (Figure 7.6).

As with van Manen's other photojournalism, such as the more recent *East Wind West Wind* (2001), *A Hundred Summers...* involved long periods immersing herself within the culture and private lives of the people she was photographing, thereby allowing her to capture images of intimacy and directness. The moments captured seem to be of the everyday, unposed and unselfconscious, or as Hans Aarsman describes it, 'a rush of images with an informal, snapshot immediacy ... pictures of Russians in their bedrooms, Russians in their showers, Russians having supper' (in

Figure 7.5 Nachtasiel. Ro Theatre Rotterdam (1998).
Photograph Bertien van Manen.

*Figure 7.6 Zavarzino Siberia. Oleg. (1992).
A Hundred Summers, A Hundred Winters*. Photograph Bertien van Manen.

Farquharson, 2003: 29). Indeed, some of the images are brutally, almost painfully, forthright and unapologetic in their direct and upfront quality.[6]

With these images part of the inspiration for Zandwijk's production, van Manen's subsequent involvement photographing the stage work intriguingly brought an artistic relationship full circle – especially when the two sets of photographs were exhibited alongside each other to accompany the production in Rotterdam. And the similarities and differences, in terms of content and ontology, between these two sets of photographs are worth exploring in detail.

Looking at these two sets of photographs side by side raises a number of questions about the differences between images captured from life and those depicting theatre (Figures 7.5–7.8). For example, it is worth thinking about the difference between the naturalism of theatre performances and the realism of the photographic image: the first of which is rooted in telling and the second in showing. Similarly, we might think of the different positions and contexts occupied by the theatre audience and the photographic viewer. An audience to a theatrical performance experiences the play within a carefully constructed, spatially and temporal determined environment. The whole production is designed to elicit particular responses and emotions. Outside of this environment it is difficult for the still photograph to capture all the emotions experienced within the theatre; instead of experiencing the theatrical event, the viewer of the still image is more inclined to see the theatrical construct.

However, it is equally the case that the reportage photograph also frames itself within a particular environment and context designed to elicit particular responses: especially in terms of asserting its integrity as photojournalism; asserting its veracity in the face of the possibility of photographic constructedness. In this context the careful labelling of each image, and manner in which the photographs are contextualised as a whole, is vital to the manner in which they are experienced as photojournalism. As Barthes formulates it, text accompanying photographs most commonly has an 'anchoring' function, enforcing an element of control and limiting the range and multiplicity of possible readings (1985: 27–30). Indeed, with reportage photographs the labelling of the image has an almost ritualistic quality, solemnly declaring the exact, actual time and place of each of the photographs. The theatre photographs, in contrast, are framed and anchored as theatre photographs, with viewers invited to read from the image to the performance. The ontological differences between the images are partly the result of these culturally constructed codes: how we use and see the photographs being significantly directed by the way we read them within these codes.

Figure 7.7 *Nachtasiel*. Ro Theatre Rotterdam (1998). Photograph Bertien van Manen.

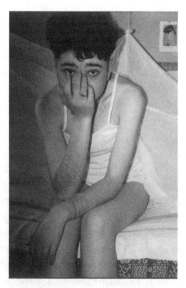

Figure 7.8 Kazan. Vlada (1992). *A Hundred Summers, A Hundred Winters*. Photograph Bertien van Manen.

It is difficult to extract the images, and their content, from these framing and contextualising codes. Nonetheless, it is interesting to ask, in terms of content alone, if the details within these images reveal the different quality of each set of photographs. As a starting point we can look at how, for example, in one of the theatre photographs four postcard or magazine images are pinned to a wall in the background (Figure 7.7). The deliberateness of these 'props', their chosen-ness as pieces of a theatrical set, seems to mark them out as constructed and posed rather than revealed. We know these details were consciously selected and put there for our interpretation and inspection; indeed, there is an entire field of theatre semiotics built around the unravelling of such clues. Yet similar details can be found in the documentary photographs. Does this reveal to us the naturalism of stage photographs, accurately constructed right down to the inclusion of these details; or does recognition of their conscious selection then lead the viewer to question the stagedness, the constructedness, of the documentary photographs?

Alternatively, looking again at these images, we might ask if it is possible to read something in the nature of the gaze between performer and camera that reveals the stagedness of the theatre photograph. Is there something about this look that is knowing, complicit, performative? Perhaps we suspect that this figure is not merely being herself, but is acting a role, a pose, a look. But equally the woman depicted in the documentary photograph is also looking straight into the camera (Figure 7.8). As reportage photography this gaze between subject and camera (and by proxy subject and viewer) speaks of honesty – both in terms of the honesty of the medium and the lack of guile in the subject. We see this as a look of unabashed frankness, the result of the intimate relationship that van Manen establishes between herself and her subjects. Her theatre photographs have sought to reproduce this candid aesthetic, but this act of reproduction casts questions back upon the original. For in the theatre photographs the look of honesty is constructed, a crafted guilelessness that partially succeeds in constructing theatre as reportage but that also partially succeeds in constructing reportage as theatre. Of course the woman in the documentary photograph is also a performer, with van Manen and her camera as her audience. Again the principal differences can be found in the differing codes by which we read the images.

Looking at the two sets of photographs, therefore, it is uncertain as to whether the documentary pictures make the theatre pictures appear more real; or if the pictures of *Nachtasiel* make us see more theatrical elements in the reportage photographs – and to an extent vice versa. As Zandwijk was inspired by van Manen's documentary photographs, and as the new set of theatre images are by the same photographer, this confusion is not surprising. But it also says more than this direct linkage in inspiration. Instead it is again apparent that all photography – whether of theatre or life – is representational at the same time as it is revelatory. We do not merely see the Russians depicted in van Manen's photojournalism, but also and always see the photographs *as* photographs. And photographs, moreover, that are in the business of representing (of *staging*) Russia to us photographically.

This is not to say that the two sets of photographs look the same, they do not. Nonetheless, to a degree the significant differences between them rest largely outside of the photographs. The documentary, authoritative power of the camera to reveal the world is itself not questioned by these images – but it is apparent that the most significant aspect to this authority is the type of perception, the way of seeing, that viewers bring to the photographs. The contextualising and framing of the photographs – in an exhibition, in a book, in a theatre programme – is almost everything: 'Tomsk railway station, 1992' makes us look in a very different manner than 'Ro Theatre Rotterdam, 1998'.

Such differences in the circumstances of the production of the photographs were also central to van Manen's own experience of the images, and with this being her only involvement photographing a theatre performance these differences were particularly consciously felt. Commenting on this, van Manen observes that 'Mostly I work close to the people and wait for what will come to me. Here I had the feeling everything came in a less spontaneous way, a more authoritarian way'.[7] Throughout the process of working on the photographs of *Nachtasiel* van Manen never lost consciousness that this was theatre. For her this prompted a different kind of sensibility: 'I was very much aware that this was theatre, because of the distance of the stage. I was not part of it.' Asked if she was aiming for her photographs of *Nachtasiel* to be read as 'real' or as 'theatrical' her answer is clear: as theatre. This difference of intent cannot, of course, be so explicit in the resulting photographs themselves, not least because it again points back to the existence of differences within the processes of production rather than in the content of the photographs. Indeed, what is particularly striking and distinctive about van Manen's theatre photographs is precisely how the performers are staring back into the camera, acknowledging the presence and gaze of the viewer. This is distinctly reminiscent of the similar gaze in van Manen's Russian images, causing us to remember that these documentary photographs are also selected: edited first from the stream of life and second from a much larger set of captured photographic moments.

As with all photography, this selection both reveals and actively constructs a particular and interpretative representation of the world.

Photo-choreography

If the revelatory authority of work by McAdams and van Manen emerges from the fact of actually having been *there*, then a very different kind of photographic revelation is called upon in the photography of Lois Greenfield. This time instead of being used to capture the world around us, photography is employed to reveal something constructed for and only visual by the camera. And yet, paradoxically, the main aesthetic remains that of revelation, if certainly no longer documentation.

Greenfield's work is unlike that of the photographers examined so far in this chapter, in that her images are not actually of performances at all. However, and this shall increasingly become a key theme in Chapter 8, photographing theatre and dance in the absence of the performance itself can be considered the norm rather than the exception. For at the same time as acknowledging the documentary motivation of much performance photography, it is also worth pointing out the odd alienation of the practice from actual performances. Originally because of technical limitations, photographers have rarely ever shown the actuality of live performance, with images much more frequently the result of scenes being re-staged for the camera in the studio. While flash technology and faster film speeds mean that stage photography is much more feasible today, for various practical and artistic reasons much theatre and dance photography continues to take place in the studio or during photo-calls.[8]

The work of Greenfield is an example of this continuing practice, although in her case it is taken to extremes as she works with dancers exclusively in the studio and has even (to a certain extent) dispensed with the dance works and choreographers altogether. Instead, Greenfield works through collaboration with dancers, improvising movement in the studio that often has no direct relationship to any external performance. The result is movement for the camera – 'photo-choreography' – that would be meaningless outside of the studio. In fact, Greenfield says that the description of her as a dance photographer 'makes me bristle' (1992: 99), and she would rather her work be considered not as documentations or handmaidens of dance but as photography as an art form in its own right. Indeed, she is less interested in interpreting choreography and more interested 'in using dancers' bodies as compositional elements to serve my own evolving artistic preoccupations' (2004).

Located outside the traditions of photo-documentation or photojournalism, to an extent the significance of what the picture is *of* in the world is downgraded. Certainly Greenfield's photographs have none of the reference to a particular time and particular place that is central to the work of Brandon, McAdams or van Manen. Ostensibly they relate to nothing outside of themselves, except perhaps to each other within the collections in which they appear. At the same time, however, these photographs are clearly *of* dance, are of dancers dancing, even if they are not of dance performances per se. Additionally, as Sontag asserts, with all photography the question of what a picture is of is always of primary importance to the viewer, as we attempt to answer this basic question before responding to any photograph in more detail (1979: 93). Although clearly not documentations of dance, therefore, as photographic images Greenfield's work will always be assessed in relation to their depiction and transformation of dance. This is particularly so as, while not reportage, her images do continue to exist within the tradition of revelatory photographic aesthetics. For while Greenfield's work would seem to be a step away from documentation, from the photographic realism of representing something that happened, this is made in the name of a deeper and more essential realism: the attempt to faithfully capture and communicate the feeling and excitement of movement.

Almost all of Greenfield's photographs are set against white backgrounds, only occasionally is the floor distinguishable and even more rarely is there a set or wall in evidence. The images appear to show people hanging in air, moments that in life would have lasted a split second. The intention is for the viewer to look at the fragment of movement represented in the photographs, to see the impossibility of stability, and ask what came a second before and what follows a second after. 'It intrigues me', writes Greenfield, 'that in 1/500th of a second I can allude to past and future moments even if these are only imaged' (1992: 116). In this manner the images are interesting embodiments of Henri Cartier-Bresson's thesis – itself a cornerstone of the revelatory aesthetic of photography – that by capturing the 'decisive moment' the still photograph can be representative of the missing whole. They also match what Antony Snowdon describes as the ambition of his theatre photography, to 'sum up a moment more than that moment' (1996: 7). Here the decisive moment seeks to lead the viewer into contemplation of movement, reading a narrative of time into the still fragment.[9]

These are, therefore, images displaying the decisive moment of extremes of movement – leaping, falling, flying, reaching – with each moment frozen, absolutely, permanently and precisely by the camera.

In one photograph, created shooting improvisational movement, three dancers seem to construct a chain of movement, as if a multiple exposure of different stages of a single jump (Figure 7.9). In the position of each of the dancers – but particularly within the composition as a whole – a lack of mobility is impossible. Our own embodied knowledge tells us that change and movement is necessary as the performers both arrive in these positions and as they descend again to the ground. In this manner we are encouraged, in some sense forced, to read a brief narrative of movement into the photograph.

Figure 7.9 Rika Okamoto, Kathy Buccellato and Camille M. Brown, Martha Graham Dance Company (1994). Photograph Lois Greenfield.

At the same time as the viewer is invited to read this narrative of movement into the photographs, we are also aware of another kind of narrative behind the photographs. Just as with photojournalism we ask about the location of the photographer in relation to any particular image, so we ask this of Greenfield's photographs. In terms of what went before and what comes after, the narrative of the images has two possible answers: one residing in the reading of the narrative of content – which is an instant in time, a frozen moment which we expand imaginatively – the other resides outside of the image and exists in the narrative of production and reproduction.

Of course, it is not always possible to know what the narrative of production behind any photograph actual was. We cannot, for example, know exactly what kind of relationship existed between van Manen and the people she photographed in post-Soviet Russia. Nonetheless we ask the question, particularly on occasions where our response to an image rests on trusting photography's revelatory authenticity. This is certainly the case with Greenfield's photographs, for as William Ewing notes of her work, 'What first must be made clear is how she does *not* work' with no darkroom manipulation, artificial supports, concealing lighting or photographic sleight of hand (in Greenfield, 1992: 14). Indeed, Greenfield presents these images as seen through the viewfinder, and one defining characteristic of many of her photographs is a hard black border, which reproduces the boundary of the negative frame within the printed photograph. The square asserts that these images are uncropped and unframed except by the lens of the camera. In many ways this device defines the ambitions of Greenfield's photographs: it affirms their revelatory status as unadulterated pictures of the world:

> I'm asking you to take my square *literally* as a boundary, not just as an arbitrary window on infinity. So now you've got the dancers butting their heads and brushing limbs against it, on hanging onto it, or being pulled off it. There's a force, as it were, that surrounds the frame that's affecting them, and it's something other than the usual conception of gravity. (1992: 116)

In doing this Greenfield constructs her studio and the camera as a new space of performance, the frame of the photograph a new stage, just as choreographers making work for video speak of the screen as a new kind of site-specific work (for example Rosenberg, 2002). Greenfield's work is similarly site-specific within the black square of her prints (Figure 7.10).

Figure 7.10 Daniel Ezralow and Ashley Roland (1988). Photograph Lois Greenfield.

In choreographing for this new space, existing somewhere between the studio and the camera, Greenfield places a striking emphasis on chance, describing elements of her work as the result of 'a chaotic situation which miraculously coheres at one particular split second' (1998: 72). This distillation within the photograph of a 'miraculous', impossible split second is the central characteristic of Greenfield's work, constructing images that seem to highlight the tension inherent within the apparently frozen balance and arrangement of the performers. These movements and positions are self-evidently not impossible to perform, but they are impossible to see without the camera: indeed, they do not exist without the camera.

For not only does Greenfield's work not refer outwards to an actual performance, nor could these dances be re-staged for an audience: the images the movements are intended to create are only visible through the act of photography. Here Greenfield utilises the ability of the camera to capture movement beyond the scope of the human eye. In this her images are not unlike Harold Edgerton's extreme stop-action photographs, including that of the coronet splash produced by the impact of a drop of milk in a saucer, which rely on acceptance of the photographic authenticity of something that without the camera would be invisible and unknown. Like Edgerton's images, even the photographer in the room cannot see them as they happen, hence Greenfield's dependence on pre-visualised chance. The resulting photographs, therefore, while not documenting the world in a conventional sense, are in the revelatory tradition of opening our eyes and showing us the world anew through the authoritative eye of the camera.

This emphasis on the revelatory quality of the images, however, only returns the viewer to think in more detail about the circumstances of their production. In particular, is the imagined narrative of movement drawn from the content of the photographs matched by the narrative of production? We know that the dancers are not on wires, but do they end up in an inelegant heap on the floor? We see the moment of perfection frozen in the photograph, but what about the moments of imperfection: the bruises, the effort, the falls and collisions. For Greenfield that the viewer is confronted with these questions has some centrality in her work, as 'the more impossible the picture looked, the more I considered it a success' (2004).

On occasion Greenfield delights in giving the explanation, which often resides not only in chance but also in sleight of hand and careful planning. In a photograph entitled *Wall/Line* (Figure 7.11), for example, she writes that 'The dancers are running sequentially headlong into the wall. The first person is held up by the pressure of the second body. The third guy has to grab the top of the wall across the width of the two bodies. The moment I shot is when the outside man, Ned, just lets go from the wall' (1998: 89). This explanation, the revelation of forethought, trial, error and effort has an authenticating, documentary impact all of its own. It acts to affirm and assert the revelatory quality of the picture, and like the magician's graceful revelation of the card trick allows the act itself to retain its power.

For despite the explanation (maybe even because of the graceful explanation), the perfection of the final image continues to enact a deliberate removal of the photograph from the processes of its construction.

Figure 7.11 Wall/Line. Ned Maluf, Christopher Batenhorst, Paula Gifford. (1994). Photograph Lois Greenfield.

As with all her work the chaos, discomfort and effort that goes into making the photographs is exiled from the resulting image. Instead they contain a calm, certainty and absoluteness that appears to brook no doubt or alternative possibilities. Greenfield herself comments on this removal of the final images from the process of their construction, commenting that 'having the dancers erase the inevitable tension in their facial expression is always one of the most difficult things about the process' (1998: 26). On another photograph she remarks that 'In the photo [the dancer] looks weightless, but every time he landed he let out a big grunt' (1998: 88). All photographs are soundless, yet this absence of sound in Greenfield's photographs can almost become their defining feature. It is an absence that makes us think back to other absences and wonder about the story, the narrative behind the construction of these perfect, frozen moments.

From revelation to representation

While a narrative of authenticity can often seem to direct responses to performance photography – responding to perceptions of disappearance – there are as many reasons to doubt the documentary fidelity of the camera as there are to be seduced by it. Apart from explicit possibilities of deliberate manipulation, the photographic image isolates time and space in a single frame in a manner intrinsically unlike the world it is representing. How can a still, silent, permanent, two-dimensional image replicate the moving, temporal, multiple, transient performance? Clearly it cannot, and the concept that any photograph might capture the 'true' appearance of performance is of limited validity. At the same time the cultural emphasis on the revelatory power of the camera to show us the world encourages us to read the photographic image in terms of their content and representation of events from the world.

The 'truth' of a photograph, however, is much more problematically located in the way in which they relate not only to the surface appearance of things but also to forces directing the cultural, social and economic meaning of things. In his essay 'Photography and Narrativity', Manuel Alvarado argues that the meaning of photographs is in part constructed through the viewer's implicit ordering of events 'before' and 'after' it. A concept Alvarado expands onto two levels: first, the order of events implied within the photography, whether 'fictional' or 'documentary'; second, the narrative or circumstances behind the production of the photography, explored through questioning 'the actual history of the production, circulation and consumption of the photograph' (2001: 151). For Alvarado, it is the discrepancies between these narratives that are crucial, and he encourages the questioning of what is *not* represented, placing political emphasis on the relationships of power and control behind the surface image. In particular, Alvarado stresses how the second level of narrative analysis – the narrative of production – is invariably suppressed in favour of the first, the narrative of content.

With the photographs examined in this chapter, discussion has repeatedly sought to identify these two different kinds of narrative through which we might approach the images. The first has been to read the narrative of content: in some sense the narrative presented to us, the one that we are invited to see and read. With photographs calling on a revelatory aesthetic this narrative of content frequently refers back to the world, claiming a veracity, authority and photographic truthfulness. Often this narrative of content seeks to elide the existence of the photographer and the photograph itself – the content, the scene, the moment is all. Within this narrative of content, therefore, we see McAdams' images as transparent records capturing canonical performances as they happened; similarly, Brandon's work provides archival evidence of ephemeral performances, valuable for the information they contain beyond their photographic status or aesthetics. Alvarado's distinction between 'fictional' and 'documentary' narrative can also be identified, whether in the narrative of movement presented to us in Greenfield's stop-action photographs or the intimate insights into personal worlds that van Manen's work seems to provide.

With such responses our analysis is significantly dependent on trust, on acceptance of the authenticating nature of the photograph and on the direct relationship between the content of the image and the world itself. However, in each instance this authenticating, revelatory power of the photography is shown – as we all have always known it to be – to be a compromised and in many ways a tenuously maintained cultural conceit. In each instance the constructed nature of the photograph begins to come to the fore, as we see how the circumstances by which they are produced, reproduced and circulated directly impacts and determines their appearance and consequently their cultural meaning. Through pragmatic or technical circumstances, what we are able to read in a photograph – what piece of the world is captured – is the result of the circumstances of production. What we see in a photograph is, additionally, also directed by what might be termed a narrative of consumption, meaning the kinds of attention that are invited and the codes of interpretation prompted through anchoring text or other framing, contextualising circumstances.

As we challenge, question and doubt photographic authenticity, so we need to read photographs in terms of the timeline and circumstances of their production and consumption, as well as in terms of their revelatory content. In these circumstances it soon becomes clear that what a performance photograph is *of* is rarely as direct or straightforward as being of a performance. Yet while the direct, documentary 'truth' of performance photographs may be fundamentally undermined, alternative possibilities of truth also start to emerge – ones no longer dependent on surface reproduction, but rather presenting emotional, cultural or aesthetic meanings and values. Chapter 8 takes this discussion forward, exploring the impact on performance photography of narratives of production, reproduction and consumption. In particular, this chapter shifts attention away from reading photography as documentary to thinking about its advertising and promotional origins, content and consumption. In this, attention moves from reading photography within a revelatory tradition to understanding it as a more explicitly interventionist and creative medium of representation.

8
Photography, Publicity and Representation

Travelling around any big city today it is almost impossible to escape photographic representations of performance. These are primarily images of performance created and distributed for publicity purposes: found on posters and billboards, on the street, in shop windows, outside theatres and on public transport. As John Berger writes

> In the cities in which we live, all of us see hundreds of publicity images every day of our lives. No other kind of image confronts us so frequently. In no other form of society in history has there been such a concentration of images, such a density of visual messages ... The publicity image belongs to the moment. We see them as we turn a page, as we turn a corner, as a vehicle passes us. (1972: 129–30)

In this performance photography is being employed in a manner little different from the use of photographs in other forms of advertising, utilising what is the medium's most technically distinct feature – its supreme reproducibility. Indeed, the ubiquity of these images makes it difficult to imagine performance disappearing, with its promotional dissemination meaning it is represented as, and even before (as shall be discussed later in this chapter), it comes into existence. Noting this public prominence and availability of performance photography, as with other forms of advertising, is significant beyond its mere ubiquity. Unlike the other forms of performance representation considered in this book (including videos and reviews), performance photography comes into our consciousness unbidden and uninvited. As Berger suggests, we see such images without effort or choice upon our part.

Indeed, Victor Burgin writes of how this availability of photographs means that 'they have no special space or time allotted to them, they are *apparently* (an important qualification) provided free of change', continuing to suggest that photography is 'largely unremarked and untheorised by those amongst whom it circulates' (2001: 66). The unremarked and untheorised circulation of promotional performance photography certainly is the case, both from the perspective of their production and their consumption (primarily by audiences, but also by academics and archivists).

Photographs presented as free in this manner – advertising photography, promotional performance photographs – encourage naïve and surface consumption. They seek, apparently without ideology, to communicate their 'content' to a potential buyer or potential audience. As a means to an end, once attracted, once bought and sold, it is tempting to think that the promotional image is replaced by the thing itself (the product, the performance) and forgotten. Viewed as 'merely' promotional, publicity photographs are indeed largely neglected as entirely secondary to the performance itself, of little evidential value and unworthy of serious study. However, this ignores the potentially lasting impact that such photographs have on audiences' reception of the performance and their role in aiding the construction of meaning from the experience. As Jan Breslauer writes:

> Production shots conclude the theater event long before the curtain descends on the final performance of a scheduled run. They beckon the audience in the first place, symbolize the event in the press, and remain as the only visual record of the work. Yet despite this power – as one of the few, if not the only, traces of what is by nature ephemeral – American theater photography exists as a practice with neither an aesthetic nor an ethos. (1987: 34)

Certainly the publicity photograph has the potential to define and direct audiences' expectations of a performance, to subsequently impact on their actual experiences and direct their memories of the event. As will be explored in this chapter, photographs are often the primary, enduring and iconic representation of events that they were constructed to promote. Indeed, it is precisely these publicity photographs that become the archival and documentary images of the future.[1] In this, Rodrigues Villeneuve is surely correct in asserting that for the performance researcher or historian all performance photography is (or rather becomes) documentary photography (1990: 32).

The meaning invested in performance photography, therefore, is not necessarily a question of the content of the image, but instead how it is used. While the photographic medium has long had advocates of its authoritative, revelatory, documentary power, this has always been accompanied by recognition of its glamorising, alluring, selling abilities. Indeed, one central ideological problem with photojournalism is the tendency to read beauty into almost any photograph. And while the importance of the documentary aesthetic in performance photography is significant, it is this representational and promotional aesthetic that is in many ways the more dominant.

As a promotional device, rather than a documentary tool, photography is a more overtly constructed, interpretative and representational medium. To explore this theme, this chapter focuses on the work of four photographers: examining Hugo Glendinning's artistic and promotional representations of Forced Entertainment performances; the choices in framing of Flemish photographer Chris van der Burght; and the computer-perfected pre-performance photography of Chris Nash. First, however, this chapter uses the relationship between the photography of Euan Myles and the productions of the Traverse Theatre, Edinburgh, as a forum through which to investigate the practice of promotional performance photography.

Advertising aesthetic

Although the documentary properties of performance photography has received most scholarly attention – with photographs considered for their value as historical records – in actuality very few photographs are constructed with documentation as their primary purpose. Exceptions include the work of somebody like Graham Brandon, working for an archival institution, but this conscious attempt to create a photographic archive of performance on any rigorous level is unusual. Photographs may be carefully stored in archives following the end of a performance run, but this is rarely their original function and many more demanding and conflicting responsibilities are required before this point. Indeed, performance photographs almost always have one immediate and economically driven function, which is their production, employment and exploitation for marketing and publicity purposes. Photographic images of performance are the most ubiquitous and effective illustrative material, used in brochures, posters and leaflets, with this function dictating when the photographs are required, what they look like and how they are taken. Indeed, the extent to which performance photography is directed and determined by the requirements – practical and ideological – of advertising cannot be underestimated.

Firstly and practically, photographs are produced according to the timescale of marketing, rather than that of performance production. Invariably this means that photographs are required long before the performance itself takes place, either at some point very early on in the rehearsals or during initial programming or conceptualisation stages. Not least for these logistical reasons, photographs taken during actual live performances are extremely rare. Instead, the majority of the photographs of theatre and dance – with the word *of* here clearly problematic – are taken before the performance actually exists in its own right.

Our cultural disposition to esteem photography according to its revelatory attributes means that such publicity photography is frequently dismissed as 'merely' advertising and consequently as unauthentic and uninteresting. From a strictly documentary point of view, the rare live performance photograph may indeed hold a greater degree of factual legitimacy. For the theatre historian its value as an authentic document of the live performance is, at least on surface levels, much greater than that of the publicity photograph. However, the amount of 'evidence' that can be read into constructed, non-documentary performance publicity photographs can be equally significant, if very different.

To explore this possibility the first example in this chapter looks at the work of Euan Myles, who primarily earns his living as an advertising photographer for companies including Irn Bru and Tennents. In addition, however, Myles also has a long-running relationship with Edinburgh's Traverse Theatre Company, creating the photographic imagery for their publicity material since the early 1990s. In this the processes behind the production and circulation of the photographs are directed by the requirements and deadlines of the marketing department, rather than the director or artistic team. As Andy Catlin, marketing manager for the Traverse, puts it: good publicity photographs are essential, as 'a really great photographic product guarantees us media coverage'.[2] Myles' photography for the Traverse, therefore, is unambiguously directed by publicity pragmatics and advertising aesthetic.

Like many theatres, the primary marketing tool at the Traverse are season brochures promoting forthcoming productions. With the theatre typically presenting new and previously unstaged plays, and with the brochures printed a long time in advance of the performances reaching the stage, the marketing department has had to solve the problem of finding suitable illustrative material for productions that do not yet exist. The plays are not, in short, available to be photographed. This is far from unusual in the performing arts, with the Traverse following a widespread solution in commissioning images that are somehow 'suitable' and could represent the content, genre or mood of forthcoming productions. Unsurprisingly for a new writing theatre, these images primarily emerge from the scripts and storylines, which do exist ahead of marketing deadlines. As Catlin describes the process, 'it's about selecting what we think is the story and trying to get something of that over in a single image'.

For example, in 2004 the Traverse produced *When the Bulbul Stopped Singing*, a play about a Palestinian lawyer and writer during the Israeli siege of Ramallah in 2002. To illustrate this production Myles found a disused quarry where the dry, cracked landscape might stand in for Palestine, dressing the location with a desk, computer and some office equipment. The image that was eventually used shows a man wearing a suit carrying a computer monitor and keyboard across the barren ground (Figure 8.1). Here the landscape juxtaposes the suit and computer to create an incongruous atmosphere, at once comic and dissolute. In the publicity material, the meaning of this image is anchored through the accompanying text, which provides the context of Palestine and themes of territorial struggle, exile and occupation. The strength of this image, however, comes from being more than simply, literally, appropriate, but instead from its evocative power that exists independently of the production and indeed of the play itself. The image invites viewers to construct an emotional narrative from the photograph that does not depend upon knowledge of the production, but which nonetheless contains something of the quality of the play and communicates suitable expectations and anticipations.

As with this example, Myles' photographs for the Traverse are cast and staged very much like mini-productions in their own right. Indeed, Myles talks of using make-up artists, costume designers and stylists, speaking of himself as a location scout and the staging as 'like decorating a room really'.[3] For Myles it is this care and attention, these details, which make his commissioned photographs demonstrably different from 'stock shots' – obtained from a photo library and used by many theatre companies for their accidental appropriateness in illustrating a production. For Myles such images are immediately identifiable as stock shots, possessing a blandness and universality that contrasts with his own consciously and deliberately constructed photographic worlds, each manufactured for a particular purpose and to communicate a particular message.

His photographs, however, rarely have any direct relationship with the production in question.[4] Recognising this, his images never seek to communicate something of a theatrical appearance. Indeed, in contrast

Figure 8.1 Advertising leaflet for *When the Bulbul Stopped Singing*. Traverse Theatre Company (2004). Photograph Euan Myles.

they are always inherently photographic (as opposed to theatrical) in their imagination and aesthetic. Characteristic of this is his use of exterior locations rather than interiors or studio shoots, his reference to traditions of fashion and style photography and the manipulation of the photographic images – as with the highly crafted and hand coloured image for *Outlying Islands* (2002), which evokes a sense of place and time in the mind of the viewer that manifests an interpretative distillation of the absent production (Figure 8.2).

In many ways Myles' photographs are very readerly images, containing material and inspiration for viewers to develop and pursue in their own imagination. Reflecting this, Myles believes that some of his images 'fail' for the slightly counter-intuitive reason that they contain too literal and too limiting a relationship with a production. One example of this is the publicity image for a play called *Olga* (2001), where the text evokes a quasi-sexual relationship between an old woman and a much younger man through reference to kissing the woman's 'cherry-red lips' – with symbolic references to fertility, freedom, youth and sexuality. The promotional photograph for the production features a grey-haired woman holding out a handful of cherries, which for Myles is 'a real cop-out image', the result of simply thinking 'oh there are cherries in the play and here is a fucking handful of cherries'. The photograph, in other words, is too literal, with a danger of resulting in a singular, *right* way of reading the image that is dependent on the play and which for the viewer, without that knowledge, is unavailable.

These publicity images, therefore, are more than merely decorative illustrations. Their first purpose is to tempt a potential audience: providing an evocation of the play, a teaser that hints at narratives, emotions and themes. Whether we consciously remember them or not, publicity images, as Berger writes, 'stimulate the imagination by way of either memory or expectation' (1972: 130). Particularly with a new writing theatre, where the audience has very little other information about a forthcoming production, the expectations and experiential promise aroused by these photographs are crucial. Accompanying these expectations are reciprocal responsibilities, with the photographs needing to be faithful to the show and provoke an appropriate horizon of expectations for the audience.

There is unfortunately little direct evidence about how audiences do in fact respond to promotional performance photography. However, one interesting insight into this question is available as a result of a complaint made to the United Kingdom's Advertising Standards Agency (ASA) about publicity for a performance by the Phoenix Dance Theatre,

Figure 8.2 Advertising leaflet for *Outlying Islands*.
Traverse Theatre Company (2002). Photograph Euan Myles.

which featured several photographs of naked dancers. The complainant, who had seen the performance, 'objected that the leaflet was misleading because the images did not resemble the actual performance'. Or, as the newspapers accurately reported the case, complained that in the performance, unlike in the photographs, the dancers were not naked (see for instance Sherwin, 2003). The ASA upheld the complaint, with their adjudication including the follow remarks:

> The advertisers said within the dance world there was a convention of using generic dance images. They said they wanted to create an image for the company as a whole. The advertisers said they wanted to draw attention to the technical strength of the dancers and to a sense of strong emotion in the performances. ... They said because they were presenting new theatrical works they did not have photographs of staged productions to use; which was a problem because venues required tour leaflets well in advance. They said box office staff were briefed that the show did not include nudity. They said the leaflet contained an accurate description of the dance works being performed. The Authority noted that one image of naked dancers appeared next to a list of performance times and dates; another photograph appeared next to reviewers' comments. Because of that, the Authority considered that readers were likely to expect that the performances contained naked dancers. It asked the advertisers to ensure that if they used similar naked images in leaflets again they made clear in the copy that performances were fully clothed. (Advertising Standards Agency, 2003)

The defence case, therefore, focused on the circumstances surrounding the production of publicity photographs and the conventions surrounding the use of such imagery. In this the defence neatly articulates the ambitions of such photographs to not necessarily communicate a strict accuracy in terms of appearance but instead a 'truth' in terms of the emotion and energy of the performance. However, regardless of the details of this particular case, or our response to the ASA's ruling, the narrative of consumption implied here is that publicity images should resemble the literal appearance of actual performances.

It is, therefore, at least anecdotally apparent that audiences do expect to see in the performance something resembling the images used in the advertising. As such the pre-performance photograph is perceived as revelatory, showing the potential purchaser what they will get if they buy this particular experience. Although the lessons here in terms of the responsibility of promoters are significant, there is also an important difference with Myles' imagery, which is not of actors or performance in the first place and that more self-evidently presents itself as a descriptive rather than revelatory medium. There is also the important factor of the longevity of Myles' relationship with the Traverse and the familiarity of his work to audiences as part of the brand identity of the theatre.

As a corporate advertising photographer Myles brings a particular quality and sensibility to his work for the Traverse, producing images that have the clean lines and perfect composition of his commercial advertising photography. His work also possesses a consistency of style that has provided the Traverse with a central element of their company branding. This has included creating cover images for the Traverse brochures, which playfully evoke the brand as something quirky, fun and irreverent, but also prestigious and of high quality (Figure 8.3). And, given the length of Myles' relationship with the theatre, it is relevant to ask to whom this brand identity belongs: as Catlin puts it 'his strong visual identity is our visual identity'. Tellingly this relationship has developed at a company aiming to attract a young and media literate audience, who will be familiar with and attracted to a promotional aesthetic that speaks more of commercial branding than subsidised theatre. In this context the sometimes almost anti-theatrical nature of Myles' photographs is appropriate to the skills and tastes of the target audience.

While absolutely not photographs *of* performance, Myles' work is very much in the business of representing the Traverse as a theatre, and therefore more broadly of representing theatre and performance to its potential audiences. In this the photographic representation of performance is very much not about a direct relationship between subject and document or subject and representation. Beyond the spatial removal of the images to the studio or found location, the temporal dissolution of the processes of production means that, not infrequently, the photographic representation is constructed prior to the existence of the performance itself. Invariably such performance photography has been neglected by scholars and devalued as insignificant, divorced, both literally and aesthetically, from actual performances. However, this neglect ignores the importance of these images in priming audiences' expectations and in coming to represent the performance not only for transient publicity purposes but also in perpetuity. For the publicity material, unlike the performance itself, is not transient, with the promotional function of these photographs joined by their role in the continuing existence and enduring representation of the performances. Indeed, these images often continue to represent the performance even

Figure 8.3 'Travel Agency'. Front cover of the Traverse Theatre Brochure (2004). Photograph Euan Myles.

when actual production photographs are available, becoming the enduring visual manifestations of each production, whether on the covers of the printed play-texts or on national and international tours and revivals. Such publicity photographs, therefore, do more than illustrate marketing material, but also come to represent and pre-present the performance in its continuing absence.

Here the classic articulation of photography as a piece of the world is overturned, and instead we can see how the world is frequently constructed and staged for photography. In this new relationship, the camera and the photographer cannot be neutral or revelatory, but instead are revealed as the creative constructors and interpreters of the world. This is something that is highlighted through the work of Myles, but which is in fact the status of much photography, including much performance photography. And this theme will be developed through each of the following examples in this chapter, exploring the potential influence that the pre-presentational role of promotional performance photography can have over both our understandings of performance and over the very production of the performances in the first place.

Advertising/documentation/art

The relationship between still photography and performance reveals a multiplicity of possible narratives of intention and reception, including the promotional, the documentary and also the artistic. One example of this diversity is found in the work of British photographer Hugo Glendinning, who takes photographs for a wide range of theatre and dance companies, with the kinds of images produced shifting according to the commissioning intentions and the circumstances of production. Central amongst Glendinning's activity is his ongoing involvement since the mid 1980s as photographic collaborator with Forced Entertainment, documenting both rehearsal processes and productions, taking images for publicity purposes and collaborating on work originating on video or photography. Glendinning's close relationship with the company is a further demonstration of the complex and thought-out approaches that Forced Entertainment have towards performance documentation. As discussed in Chapter 4, this includes both pragmatic and artistic attitudes to performance representation, both of which are also in evidence in Glendinning's photography for the company.

The pragmatic attitude to representation is revealed in the production of publicity material, where Glendinning states the objective is always unambiguously

> to make something that can sell a show ... I just do whatever works for me and assume that it will look good on a poster ... I just try and make it a good photograph. You know, firstly. Because most shows can carry almost any photograph.[5]

As already explored though the work of Euan Myles, the general objective here is substantially that of many performing arts companies: the production of a strong visual image that catches the eye and is effective in selling the production. This is always the primary factor in all promotional performance photography, where a 'good' photograph is evaluated according to its ability to fulfil this function. This is not to neglect the sheer range of different kinds of imagery used on the publicity material by Forced Entertainment, where (more than with Myles' consistency of style) the imagery varies dramatically as it seeks to be evocatively appropriate for each different production. However, Glendinning's statement that most shows 'can carry almost any photograph' emphasises how any evocative relationship to the production is entirely secondary to its function within the publicity material and marketing campaign – is secondary, that is, to the photograph's evocative relationship with the viewer.

For the images to work as publicity, therefore, they must construct and contain their own narrative and meaning within the frame of the advertisement. The advertising flyer for *Emanuelle Enchanted* (1992), for example, constructs an ethereal sense of calm within movement not through reference to the absent performance, but instead through qualities contained within the image as an image (Figure 8.4).

While we cannot compare the imagery of this advertisement to the now absent live performance, thereby exploring what particular selections and interpretations it enacts, it is possible to compare this leaflet with the uncropped version of the same photograph held in the company's archives. With a much wider framing, capturing the full stage width and providing greater background detail, this act of comparison immediately, if somewhat perversely, invests this 'original' picture with qualities of completeness and authenticity in comparison to the advertising image. For example, the tight cropping of the publicity image focuses our attention solely on the drawing of the curtain and the running man. The dominant perception is one of revealing, which the text anchoring the image prompts us to consider in terms of the uncovering of beauty in ordinary things. This fairly directed reading is in contrast to the uncropped image, which is messier and less singular in its focus, presenting several different and distinct possibilities in interpretation. Although the running man and curtain are still central, here an expanse of stage space, scattered papers, chairs, unadorned stage flats and fully laden clothes rails are also visible. All contribute to the overall balance and meaning of the photograph, complicating our reading of the image.

However, as important as this difference in content, is the contrasting approach that is encouraged by the context of the two images. The publicity image is read in and of itself, with its narrative and meaning contained within the advertising media and the framing and anchoring effect of the text. So rather than the stage being cropped out, for the viewer it simply does not exist. Equally, we do not ask what is to the running man's left and right, or ask why he is running in any literal sense, but instead interpret the self-contained world constructed within the advertisement. In contrast the uncropped image is framed and anchored in a very different way. Its factual labelling demarks it as an archive photograph, directing viewers to read the image in terms of its representation of an original event (an impulse which may in fact be mistaken, with the photograph potentially produced from a photo-call or rehearsal rather than an actual performance). Additionally, the absent space and

Figure 8.4 Advertising leaflet for *Emanuelle Enchanted*.
Forced Entertainment (1992). Photograph Hugo Glendinning.

time are now continuous: what is to the right and left of the running man, what happens before and after the moment depicted, are now historical, factual (if unanswerable) questions. Instead of presenting a self-contained world, as an archival image we know there is a narrative that is not available to us through the photograph alone.

Some of the key differences between what might be termed documentary and publicity photographs, therefore, are constructed more through how we use and read them than through their actual content. The framing and contextualising of any photograph – as publicity, as documentation – significantly directs how we respond to the image. This is also the case with some of the photographs by Glendinning of Forced Entertainment's *Club of No Regrets* (1993), where a powerful mythology has been invested in the images that almost enforces a particular ideologically directed reading (Figure 8.5). In *Certain Fragments* Tim Etchells describes the circumstances in which these photographs were produced:

> When Hugo saw *Club of No Regrets* for the first time we were about a month or more away from completing it. And the show was really fairly out of control. There were scenes for which the performers were parcel-taped to chairs, and the scenes in which large amounts of leaves, water (for rain), talcum powder (for smoke) and fake blood were thrown around the stage. ... And after watching a run-through of the material Hugo and I talked a bit about how he might photograph the work and he said that he'd 'like to take the pictures in the equivalent way to the way we were working theatrically', and I asked 'What's that then?' and Hugo said he was going to photograph the piece without looking through the camera.
>
> He shot without looking – a flash gun in one hand and the camera in the other ... camera at arms length, held above his head, thrust right into the middle of scenes, never certain what he'd get. But a part of the action. (1999: 110)

As with Etchells' descriptions of Forced Entertainment's artistic acts of self-representation (discussed in Chapter 4), the interest here is in qualities of fragmentation, trace, dissolution, erosion of certainty and partiality. The resulting photographs can also be seen as foregrounding the postmodern stress on the selectivity of the camera, rather than the medium's revelatory power. The images certainly do have a raw power, an elusiveness – the 'photographs seem not so much to contain the event as to hunt it, always losing' (Etchells, 1999: 110) – that speaks

Figure 8.5(a)–(d) Photographs from the 'blind shoot' of *Club of No Regrets*. Forced Entertainment (1993). Photographs Hugo Glendinning.

both of the absent liveness of the performance and the qualities of this particular production.

As noted by Etchells, the photographs also, intriguingly, possess a narrative of production that begins to match the narrative of the performance itself. However, as with my concern over the way in which the fragmented self-representations of Forced Entertainment's work actually begins to close interpretations and construct a singular authoritative reading, the same can be said of these photographs. Or perhaps more accurately, less with the photographs themselves than with their perceived ontology and the mythology surrounding their construction, which is seductive and poetical but at the same time too easily accepted in a manner that ironically leaves doubt, questioning and uncertainty at the door. Tellingly, while Glendinning agrees with Etchells' description of the process behind the *Club of No Regrets* images,[6] he also describes it as primarily a photographic approach that he has also employed elsewhere. The intention, Glendinning explains, being 'to really fuck up my place' and challenge the way in which the question of how to frame a photograph can tyrannise a photographer:

> So I'm just going to let the equipment do the job and I'm not going to do it at all ... not looking through the camera, not focusing, you know, just scattergun. ... I'm conscious of wherever I put anything in the frame, and if I consciously do it I get irritated by myself. So it was a way of re-framing anything in an arbitrary way. That's one sort of valuable thing. And you just get chance things. ... I did it with Fiona Shaw in *Medea* for a West End production. You know, just holding a camera and not trying to make anything, not trying to make pictures, just trying to be with the performance and move with it in a fairly organic way.

Crucially, and pragmatically, Glendinning and Forced Entertainment also produced 'straighter', conventionally (i.e. 'well') framed, colour photographs of *Club of No Regrets*, the existence of which re-opens possible readings of the production, allowing the construction of new perspectives and interpretations. A question remains as to which of these sets of photographs 'best' represents the production, a question that depends on what you want the representations for and how you are going to use them. The slippage of performance photography between functions of publicity, documentary (of surface appearance or emotional meaning) and artistic self-representation potentially exists in the same image, our ways of seeing them transformed through their anchoring and framing within established narratives of production and codes of reception.

Artistic authorship

The discussion of the work of performance photography raises knotty questions about the 'authorship' of the photographic representation and the relationship between the photographic style and form and that of the performance depicted. For here the performance is reinterpreted and represented through the transformative imagination of another artist. It is possible, therefore, to see the subject or content of the photographs as the work of one set of artists, with the style and aesthetics of the photograph as the work of another. Alternatively, is the form of the photograph inextricably directed by the form of the performance, with the camera employed as a relaying tool or implement? Working in close collaboration with the artists concerned, Dona Ann McAdams, for example, seeks to efface the presence of the camera as an interpretative, artistic medium in its own right. Indeed, in declaring the transparency of her photographs, she even goes as far as to claim that 'This is not my work' (1996: 8). For Lois Greenfield the reverse is the case, with her photographs very much presented as works of art in their own right, constructed through creative engagement with her subject. In fact with both these examples, as with the work of Myles and Glendinning, the relationship is more complex than this with the photographs the result of a collaborative engagement between the media of photography and performance.

This question about the relationship between the style and form of the photographic representation and that of the performance depicted is relevant to all performance photography. One example through which to explore this in more detail here draws on the work of theatre and dance photographer Chris Van der Burght and in particular his images of a production by Het Zuidelijk Toneel of Marguerite Duras' play *India Song* (1998), directed by Ivo van Hove. (This example also refers forwards to Part IV, where this production of *India Song* provides the central illustrations in discussions of performance reviews.) Van der Burght's work also continues another theme in these chapters concerning the choices made in the framing and composition of the performance photograph. Here the striking choices of framing made by Van der Burght disobey standard rules of composition, making explicit his use of the camera as an overtly interpretative eye. For Van der Burght these choices are partly instinctive, the result of suddenly seeing and grabbing a momentary image, indeed he describes such flashes of images as central to our way of seeing, as 'the only things we recognise'.[7] The result is that his images are frequently easily identifiable as the product of his distinctive gaze; yet at the same time also seem to achieve a representation appropriate to the particular nature of the production in question.

Each of Van der Burght's photographs of *India Song* is printed to include a hard black border surrounding the edge of the image. As with the work of Greenfield, inclusion of this border asserts the 'authenticity' of the photographs in their uncropped presentation of what was seen through the viewfinder. In this context, however, the inclusion of this border also serves to deconstruct this statement of revelatory authority as it immediately demonstrates the conscious choice and selection behind the photographic composition. One photograph, for example, centres on the joined hands of two performers, as one leads the other across the stage, which is visible but out of focus in the background (Figure 8.6). Their bodies are brutally severed by the photographic frame as they appear and disappear at the edges of the image, this movement across the frame of the photograph evoking a sense of movement across the stage itself. In its composition this image makes its status as a photograph clear, as the viewer cannot fail to notice the act of selection. The inclusion of the negative border around the image now draws attention to the photograph as deliberately composed and constructed, rather than a transparent window onto a found piece of the world.

With this photograph the viewer is required to consider why this selection has been made, and, rather than passively seeing the photograph, begin to actively read the image as an interpretative representation that constructs the appearance and meaning of its absent subject. While some of his other photographs of *India Song* are composed more conventionally, Van der Burght exclusively avoids the head and torso shot, rarely focusing on the actors as televisual 'talking heads' but instead on bodies and bits of bodies, and on a clearly (if only partially) present stage. In focusing only on bits of things, his images make visible the fact that they do not record everything; the gaps and omissions rendering the medium evident, and ensuring remembrance that the theatrical event itself is absent. On this Van der Burght intriguingly states that 'To make it possible for me to work as a theater photographer, I chose to take the theatrical as reality, a reality without physical borders. This reality doesn't stop at the edge of that stage.' Similarly, his photographs do not stop at the edge of the frame, but instead continue through our imaginations into a constructed remembering of the theatre.

Van der Burght's photographs encourage us to think of the absent performance event partly by continually reminding us that they are not

Figure 8.6 *India Song* (Chris Nietvelt and Ramsey Nasr).
Het Zuidelijk Toneel (1998). Photograph Chris Van der Burght.

complete. The images fail to contain all their reference and relevance within the frame itself, forcing the viewer to read narrative and context, and time and movement, into the image, but also to think out beyond the image. As importantly, however, they evoke the absent performances through the theatrical qualities present in their own form – there is a stagy dynamism to the images that speaks of movement, process and theatre – they are images that, as French theatre photographer Claude Bricage demands, dare to stage themselves (in Villeneuve, 1990: 36). One image, for example, strikingly severs the head of the central character, a decision that in turn draws our heightened and interpretative attention to the face and gaze in the background that is visible (Figure 8.7).

Yet at the same time as possessing a powerful and self-consciously photographic representational style of their own, Van der Burght's photographs of *India Song* are also strikingly appropriate to the nature of the production they depict. The obscured faces in the photographs, the leading hands, the trailing lines of costume and gesture, the indistinct rendering of the stage in saturated colours all construct a sense of insubstantialness that replicates the style of a production where the performers did not speak 'live' on stage but instead drifted – present but somehow only partially so – as the dialogue and stage directions were relayed through speakers mounted above their heads. On stage the separation of voice from actor rendered the characters partially absent from their own bodies. In the composition of these images Van der Burght has found a photographic equivalent to this intrinsically theatrical device.

Bearing this in mind, it is intriguing to look at other examples of Van der Burght's work, where similar stylistic motifs are in evidence. For example, he frequently takes photographs for choreographer Meg Stuart and Damaged Goods, including *Visitors Only* (2003). Looking at his photographs of this production, a number of aspects are immediately familiar, including the inclusion of the border around the images and the use of an often theatrical and self-performative framing, used to bisect performers, sets and action (Figure 8.8). Here the impact and meaning of these techniques is subtly modified by the content of the images, yet seeing this style transferred to another performance prompts the question as to the authorship of the photographic representation. Again we need to ask to what extent is the aesthetic drawn from the particular production or the result of the distinct gaze of the photographer.

The recognisable quality or authorship of a Van der Burght photograph does not result in all the productions he photographs looking the same. Nonetheless, the photographs significantly direct our vision and

Figure 8.7 *India Song* (Chris Nietvelt and Bart Slegers).
Het Zuidelijk Toneel (1998). Photograph Chris Van der Burght.

Figure 8.8 Visitors Only. Damaged Goods (2003).
Photograph Chris Van der Burght.

conceptualisation of performances, which are only visible to us through the particular form and style of the photography. The appropriateness of Van der Burght's work for both *Visitors Only* and *India Song*, might therefore be less that he has adopted his approach for each particular performance, but rather that he has gravitated towards taking photographs for practitioners sympathetic with his own aesthetic and way of seeing. Either way, it appears that the success of the relationship is the result of the photographic approach being apposite in form and style for the particular performances in question.

That the representational style of a photograph should match that of the performance is in many ways the ideal. It is certainly something that has been observed in the relationships between photography and dance, where particular choreography has come to be identified through particular styles or aesthetics of photography.[8] So, for example, Fergus Early writes of dance photography that Anthony Crickmay is a photographer:

> Whose individual 'eye' has done most to contribute to contemporary tastes in dance ... His most characteristic work is shot in the studio against a pure background, just the dancer's figure, streamlined in

white light, soaring, floating, etherealised. It is an aesthetic that is not so very far removed from that of a TV commercial, packaging the dancer's body in an alluring but trustful manner, yet it is an aesthetic that is in tune with some of the more dominant trends in both ballet and contemporary dance. (1984: 4)

Significantly, Early was writing in 1984, and tastes and trends in contemporary dance have changed dramatically since then. In part this has been marked by a shift away from the ethereal image of dance as embodied in beautiful bodies, and towards a perspective that looks at the everyday movements and everyday bodies of dance-theatre – as demonstrated in the work represented in the next example explored in this chapter. Also relevant, however, is the extent to which this movement has been motivated in reaction against the ethereal choreography of the 1980s, or away from the etherealised imagery of Crickmay. In other words, was the style – the manifestation of the style, the style made visible – the result of the choreography or the result of the photography?

There is no singular or stable answer here. Instead it is more profitable to see a symbiotic relationship between the photographer and the choreography, with each depending on the other – the choreographer needing the photography to represent the work beyond the performance; the photographer needing the choreography as the material, the content of their work. What is apparent is that the camera does not passively capture or reveal these performances; instead the photographer is engaged in a careful and considered act of representation. The choices and selections – in terms of technical decisions, in terms of objectives, in terms of what to shoot and how to shoot it – are central and determining to what the photograph, and through that the performance itself, looks like. Indeed, to a great extent we cannot see except through the vision and interpretation presented to us by the photographers.

The computer as studio

Photography, therefore, does not passively record the style of the performance. Instead, in many ways it is the transformative act of photography that makes the work of the performance artist visible. A revealing example of this is found in the work of British photographer Chris Nash, whose most characteristic images demonstrate an almost complete fracture with any enduring revelatory aesthetics, as instead he employs the camera as a much more explicitly creative tool. Considering first his painterly style of representation, and later his creative collaboration with the artists whose work he represents, Nash's photography is the final detailed case exploration in this chapter.

While Nash has been taking photographs of dance for more than twenty years, like many of the photographers whose work has been explored here, his images are often not *of* actual performances at all. His current activity is almost exclusively studio-based, mostly commissioned by dance companies and choreographers for promotional purposes. Characteristic examples of this include Nash's work with choreographer Lea Anderson, such as the publicity image he created for *The Featherstonehaughs Draw on the Sketchbooks of Egon Schiele* (1999), a piece to be based on the portraits of the Austrian painter but where this image was created long before the work went into rehearsal (Figure 8.9). There is a jerky, awkward physicality to the image, drawing our attention in deeper and for longer as we find an echo or residue of movement in the stillness. Close inspection, however, also poses a number of problems for the viewer: what happens to the dancer's right leg as it passes behind his left knee? What is his relationship with the ground? And, if this image suggests movement, in exactly which direction would he be moving?

The difficulty of answering these questions holds our attention further, for although we may begin to suspect photographic trickery the strength of the image stops us from dismissing either the photograph or the movement it communicates as 'fake'. For although faked in the conventional sense of the word – this is not a photograph that shows us a piece of the world in a revelatory fashion – the image does not seem false in any reductive manner. Instead, it is a self-consistent realisation of particular photographic ideas and interests, something that Nash's narration of the production process behind the image makes clear:

> The visual influence in the photograph is fairly straightforward, with the use of Photoshop to produce the writhing edges and figurative distortions that typify Schiele's work. However, I wanted to also replicate the way that his full length figures are evidently posed, often lying down, yet still appear to be twisting and writhing – a factor that had drawn Lea to the subject matter in the first place. To do this I elected to pose the dancers (rather than try to capture a particular movement) in a way that suggested movement. We used chairs for the dancers to balance on (subsequently taken away in Photoshop), enabling apparently gravity-defying positions.[9]

This use of the computer, and openness about it, is typical of Nash, whose work frequently uses montage, the rotation of the frame, digital

Figure 8.9 *The Featherstonehaughs Draw on the Sketchbooks of Egon Schiele.*
The Featherstonehaughs (1999). Photograph Chris Nash.

manipulation, bright and contrasting colours and other graphic interventions made on the computer to achieve particular ends. The ambition with these techniques is not to communicate the empirical appearance of the dance (which in fact does not exist anyway), but instead to represent a more essential spirit, excitement and distillation of the dance to be. Nash's approach is creative, interventionist and painterly, something that can also be seen in his use of computer inserted colours and backgrounds to communicate particular emotions and moods to the viewer. In this careful construction and staging of photographic space Nash articulates his work as being 'just like theatre. The frame of the photograph is like the proscenium arch. You know you're being shown things, but it's how it's shown and what is shown that's interesting.'

The images, therefore, are constructed theatrical experiences in their own right, in which Nash does not hesitate to use interventionist techniques that often go against conventional use of the camera to show perfect surface appearances of how things 'really' are. By employing these methods Nash seeks to create (rather than capture) an image that communicates the character of the performers and choreography he is representing. Another example of this is Nash's publicity photograph for Yolande Snaith's *Blind Faith* (1997, Figure 8.10), of which he says:

> Yolande wanted to portray a powerful women who could inspire acts of blind faith such as levitation, so we set up this shot so that I could manipulate the image in Photoshop to give the appearance of a floating body. Does it matter that the camera has lied? The camera always lies! Or rather, the truth is relative and depends on where your viewpoint is. If we can agree that a photographer plays an active part in the creating of a photograph, then is he or she not already manipulating the information contained in the frame?

Nash is certainly correct here. Our cultural faith in photographic authenticity has always been based on some kind of willing suspension of disbelief, and in many ways we always have recognised that the camera is a tool that we use in an interpretative and selective manner. That is not to say, however, that the revelatory aesthetic is removed entirely, either from Nash's work or from the way in which we view photography more widely. Instead there continues to exist an interesting and tangled relationship between choice and chance, between conscious selection and construction and accidental revelation.

Figure 8.10   Blind Faith (Yolande Snaith and Paul Clayden).
Yolande Snaith Theatre Dance (1991). Photograph Chris Nash.

As with other photographers discussed in this chapter, there exists a close correlation between the kind of photographs Nash produces and the kind of performances he depicts, with Nash particularly associated with dance-theatre and postmodern dance productions. Here the use of digital manipulation is very appropriate to the particular companies and choreography with which Nash works, with his approach to photography matching the choreographers' approach to their work and sharing a suspicion of 'beauty' and athleticism in choreography for its own sake. In the choreography there is an interest in deconstructing dance, taking movements from the everyday or martial arts and in the crossing points between dance and visual art, film and literature. Similarly, Nash's photography is interested in deconstructing the image (and its cultural claims to authenticity), blurring the boundaries of what is considered photography and blending it with other media including drawing, painting and digital art. These shared concerns produce a sympathetic meeting between the style of the photography and the choreography. The surprise is that to achieve this Nash's photographs have to be more interventionist, less 'realistic' as they undermine surface fidelity to communicate a constructed representation of the choreography. Another key factor here is that to make this possible Nash has developed highly creative and collaborative relationships with the artists concerned.

Creative collaboration

It is noticeable that the most interesting performance photography is frequently the result of a close collaborative relationship between photographer and artist. This is something acknowledged, not least, by the artists themselves, who recognise the creative input of the photographer in making their work visible. Karen Finley, for example writes of McAdams' photography: 'I learn something from seeing her work, because the image speaks to what the piece is about. It makes my work real' (in McAdams, 1996: 8). Tim Etchells has made similar comments on the relationship between the work of Forced Entertainment and Glendinning's photography, writing of the process of creating *Club of No Regrets* in particular that seeing the photographs 'was also a really vital part of the making of this piece – because in seeing the photographs we could see, for the first time, what exactly we were doing' (1999: 110). On occasions such collaborative relationships can go further, with the meeting between photographer and artist becoming a place of constructive, creative engagement. This degree of creative collaboration is what emerges, perhaps surprisingly, from the marketing and publicity driven processes behind Nash's dance photography.

Employed to construct pre-performance publicity photographs the timescales of marketing and programming mean that Nash, as with previous examples, is typically working long before the productions in question are in full rehearsal, let alone ready for the stage. Like Myles' photography for the Traverse Theatre, therefore, Nash produces photographs intended to represent performances before they exist in fact. Unlike Myles, however, whose contact is almost exclusively with the marketing team in a fairly formal client relationship, Nash frequently works in very close co-operation with the artists themselves, producing photographs that are highly collaborative and mutually inspirational. In creating his photographs, Nash works with the choreographers to anticipate and articulate visually what the feeling and look of an embryonic dance might be. With no pre-existing or original script, these dances are often barely past the conceptualisation stage. Together with the choreographer, Nash seeks to find a way of putting sometimes fragmentary and raw ideas into a picture that possesses some of the emotion and meaning that the choreographers hope the audience will draw from the final work. The results are described by Nash as 'my interpretation plus the choreographer's interpretation of what the piece is about'.

In Nash's work, moreover, it becomes very apparent that the close collaboration with the artist at this pre-performance, pre-rehearsal stage of the creative process can produce important cross-fertilisation of ideas and concepts. Indeed, for Lea Anderson the photo-sessions are a conscious opportunity to work with Nash in the studio to explore ideas (such as the shapes from Schiele's paintings) that would form the central concept of a new dance piece. Similarly Javier de Frutos, who has worked with Nash for over eight years, describes how the photographs are created in a meeting place between the pragmatism of the marketing process and creative collaboration and engagement:

> Publicity shots are requested by venues so early in the creative process that most of the time rehearsals have not even started. Yet, I always treasure the photo-shoots as a playtime that allows me to test some of my research into the first visual collage of ideas later to be staged. (In Nash, 2001: 42)

The intriguing possibility, therefore, is that the reversal of the expected chronological relationship between performance and photography means that the act of constructing an image to represent the work in promotional campaigns has the potential to aid and even direct the act of creating the performance itself. When asked about this in interview, Nash comments:

> Lots of choreographers have said that the photo-shoot is a chance to crystallise some of their ideas and to try things out. And also, seeing them as a photograph you can then get other people's reactions to them and it's another way at looking at your ideas. ... Because we go in with some basic ideas, and then improvise around those themes, the improvisation is quite useful to the choreographers as well. And usually those sessions are right at the beginning of the rehearsal

process, so they are the first chance for the choreographer to actually try some things out.

The potential for this to occur is largely accidental, resulting from the requirement for publicity photographs long before the existence of the show itself. This accidental nature is also slightly illicit, with the primary motivating factor being marketing, not art, meaning that public and critical recognition of the extent of this mutually inspirational relationship is often hedged and concealed. For various reasons neither side in the relationship may feel open or able to explore it freely. However, as seen with Nash it certainly exists, with another example involving the work of Glendinning, as described by Matthew Bourne when talking about his choreography for *Swan Lake* (1995):

> A lot of those movements came about because Adam and I had to do a photo-shoot with Hugo Glendinning long before rehearsals and we hadn't any movements at that stage. All we had was a costume. It was easy to call upon some of those images, to play around with arms, looking through features, shielding the face. Having come up with these photographic images, we had a starting-point when we went into the rehearsal studio. (In Macaulay, 2000: 241)

Talking to Glendinning it is clear that this kind of creative collaboration between photographer and director or choreographer within the processes of publicity photography is far from a unique occurrence. Other instances include occasions in Glendinning's ongoing relationship with Forced Entertainment. Indeed, Etchells describes how with *First Night* (2001) the whole show blossomed from the publicity photo-shoot, which occurred right at the beginning of the rehearsal process and fed directly into the look and style of the piece. For Etchells, working with Glendinning on the publicity photographs can provide a clue as to where the process is going to go, although he also adds there is no structured way of that happening or of looking for that to happen.[10] It is worth stressing that such relationships are not planed nor inevitable and also warrant a degree of sensitivity in discussing. Nonetheless, the positioning of photographic engagement with performance right at the beginning of the production process means that this kind of creative collaboration and inspiration is probably far more frequent than either the artist or photographer are aware themselves.[11]

As we have seen, Nash works with choreographers to make visible the ideas and emotions that they bring to him. His work shows imagined happenings and visions that are staged for and realised by interventionist, representational photography. These can then become kinds of templates for the performances yet-to-come, as the choreographer goes on to construct a dance work that is sometimes directly inspired by the vision staged for the camera. The image therefore possesses a complex pre-representational relationship with the work itself. Indeed, Yolande Snaith writes of how in performance one particular section of *Blind Faith* 'was the closest we came to realising the photographic image' (in Nash, 2001: 40), thereby encapsulating the complex pre-representation relationship between photograph and performance, between origin and depiction.

A final example of Nash's work can be seen as exactly this kind of staging and perfecting of the choreographic vision. Another Featherstonehaugh image, taken to publicise *Immaculate Conception* (1992), shows two groups of dancers: below a darkly dressed group huddled around a prone figure; above white-clothed figures reaching down from a cloud (Figure 8.11). There are hints in this image towards Michelangelo and a renaissance inspiration, along with a wry humour in the combination of kitsch and pathos. Again this photograph sees the pre-performance realisation of a choreographic concept, this time of a two-storey platform as a performance space. And again this image has been completed on the computer, with the two levels of the performance space perfected – no scaffolding or supporting wires, necessary in performance, intrude into this realisation of the vision.

In this and his other photographs Nash removes the physical limitations of live performance itself, obliterating the possibility of errors, eradicating the staginess of stage sets and thereby idealising and perfecting the anticipated theatrical illusion. Although not *of* performance, therefore, Nash's work is not divorced or isolated from the envisioning of performance. And, indeed, after the work has been staged, after the run is over, the photographs remain and assume a new documentary function – recording if not the performance in fact then instead the performance of the imagination.

The age of digital reproduction

Perceived as a fundamentally revelatory medium, our responses to photography are directed by its qualities of documentation and reportage: we look at the photographic image only and always in relation to what it shows us of the world. At the same time, however, photography,

Figure 8.11   Immaculate Conception. The Featherstonehaughs (1992). Photograph Chris Nash.

whether ostensibly news, reportage, advertising or domestic, also invites viewers to engage with the image on its own terms, as a work and a statement in its own right. While the passing of time means that all performance photography becomes documentary in function, our cultural relationship with photographs means that we always also assess them in terms of their aesthetic value. Certainly all photography has the ability to resonate with a presence, beauty and meaning of its own, independent from its subject or original purpose.

Indeed, photographs of performance frequently engage us as great works in their own terms. As Joel Schechter writes, 'photographs are the only visual record of past performance that has ever been acclaimed as great, even immortal' (1987: 4). This is in stark contrast to video recordings of performance, which (as discussed in Part II) rarely make great videos. One explanation for this is that while video is itself a medium of performance, the greater distance between performance and photography necessitates greater intervention and artistic interpretation. It is this inability of the still photograph to capture a performance easily or completely that renders attempts to do so more significant, with each choice and act of selection communicating an interpretation of what was considered significant, meaningful and valuable in the performance. This is the case even with the constructed, conceptual, 'fake' publicity photograph, which although not *of* performance in any direct sense continues to hold different but still extremely valuable information to the theatre and dance researcher.

It is, of course, not always possible to tell from a photograph itself exactly what the circumstances of its production were: was it taken during a performance, rehearsal or photo-call; or in the studio; has it been posed for the camera or caught candidly; has it been manually or digitally edited? With publicity photographs such questions are particularly relevant: frequently taken before the performance exists in its own right, the posed, touched-up, studio promotional photograph holds little direct factual information about the performance itself. Consequently, from a strict historical interest in the recording of literal moments of a performance, the promotional image presents a problem to be ignored or overcome as a distortion rather than a true record of the history of the stage.

This, however, is to elide not only the nature of photography as a medium but also the economics of production that shape and direct performance itself. As the examples in this and the previous chapter have explored, all photography is interpretative, constructing the world at the same time as it reveals it. Through necessity and the pragmatic concerns of marketing and publicity, promotional photography has had to more explicitly forgo a revelatory aesthetic, instead employing the medium in a far more self-consciously representational and painterly manner. As a result, it is possible to see how pre-performance photography not only subverts the assumed temporal relationship between the document and the thing itself, but also destabilises any aesthetics of photographic revelation and authenticity.

For in addition to faith in the camera's chemical veracity, the aesthetics of revelation and the photographic documentation of the 'real' also depend upon faith in the existence of a single and recordable real. Today we are no longer in what Walter Benjamin termed the age of mechanical reproduction, but instead in an age of digital reproduction. Photography as a medium is no longer the result of a mechanical and chemical process, and the impact of this on photographic aesthetics is huge. The perception of photography as a mechanical, chemical form existed in a largely singular, modernist and hierarchical world view. In contrast digital photography enacts the fluidity of the world, demonstrating the centrality of selection, of editing, of point of view, of change and of transience. The digital camera instills photographs with a much more postmodern ontology, particularly so in relation to performance photography where the photograph itself is often of a constructed rather than a found or revealed world.

Perhaps this should be regarded as nothing inherently new, with claims to ontological differences between mechanical and digital photography calling upon too essentialistic a binary. Indeed, distortion and transformation is inherent in all acts of photography, with differences between pre-shutter construction of the image (posing the subject) not inherently different from post-shutter or digital manipulation. As Hella Beloff writes, we should recognise that 'from the start, both practitioners and theorists of photography have understood that photographs can lie. It is working out why and when and how they do it, while at the same time seeing that they also show a truth, that is so difficult' (1985: 19). And equally the digital camera does continue to partially exist and operate alongside a continuing faith in a revelatory aesthetics – one world view has not supplanted the other. Indeed, it is the tension between the way that the camera constructs and at the same time reveals a truth that makes the medium today so vibrant. The result is that the depth and richness of photography has only increased through digital technologies. In the age of digital reproduction, the power of performance photographs continues to be in part the result of the camera's revelatory ability to capture the world, but is as much the result of its representational abilities to construct the world – the result, that is, of choice, craft, technique, manipulation and intervention.

Part IV

9
Reviewing Performance

As a form of performance documentation, writing is self-evidently different from either of the media, still photography or video, examined in the previous sections. Writing does not bear a direct mimetic relationship with the world and cannot claim mechanical objectivity, audio/visual completeness or documentary transparency. However compromised such assertions may be in relation to photography and video, writing about performance is evidently much more inherently transformative. As Auslander observes, 'Written descriptions and drawings or paintings of performance are not direct transcriptions through which we can access the performance itself, as aural and visual recording media are' (1999: 52). In which case, Auslander suggests, although audio/visual media may transform performance in the process of recording it, they nonetheless provide a directness of knowledge that non-mimetic forms do not allow.

This is undoubtedly the case, although it does also to an extent downgrade the presentational qualities of writing. Additionally, the transformative impact of writings about performance can be viewed as an opportunity for analysis rather than a problem to be overcome. Choices in representation make accessible and concrete perceptions of performance, thereby making performance manifest as a cultural phenomenon. In other words, it is exactly the transformative nature of writing that is of interest here, using it as a prism through which to explore what is valued, remembered and understood about the absent live performance.

Under the broad theme of the written representation of live performance there is the potential to explore a vast range of possibilities: including letters, diary entries, internet discussion forums, reviews and other written criticism and so forth. Also of particular interest is academic writing and discourse, such as explored in Chapter 3, where there has been important and extensive debate regarding attempts to document performance and write the live in language (such as Banes, 1994; Melrose, 1994 and Pollock, 1998 amongst many others). Indeed, in her essay 'Performing Writing', Della Pollock seeks to identify the nature of 'performativity' in writing, suggesting that 'writing as *doing* displaces writing as meaning; writing becomes meaningful in the material, dis/continuous act of writing' (1998: 75). While it is undoubtedly to decontextualise and displace Pollock's primary intention, it is possible to see this description as better suiting the review (and the processes by which reviews are written, published and consumed) rather than the academic discourse. For the review, like performance, is a form of *doing* or a kind of action in its own right. Indeed, much of the practice of reviewing is about the process of writing itself, about negotiating form, content and meaning within the context of both the production (and consumption) of reviews and the production (and consumption) of performances.

Consideration of the review has often, and importantly, looked at the social and artistic power of the critic (see for example Booth, 1991; Elsom, 1981; Féral, 2000 and Shrum, 1996). While keeping this vital context in mind, the intention here is to focus on reviewing as a presentational (and representational) act, examining what aspects of theatre and dance are presented and considering how critics construct their experience of the event through language. In this context the review has often been seen it as an inherently compromised form, whether by its origins and circumstances of production, its mainstream focus or its lack of greater ambition or self-reflexivity. At the same time, however, the publication of reviews in mass-circulation newspapers provides a prominence that makes them by far the most widely circulated and consumed form of written performance representation. Additionally, more than the other instances of written discourse, reviews are *about* live performances, with one of its primary functions being to present the performance to readers who were not there.

The unsystematic but custom-defined publication of reviews by newspapers also means that they form a kind of automatic or embedded documentation of performance history as it happens. In this reviews are somewhat like the performance publicity photograph. Both are representations of performance that are constructed according to convention, expectation and most significantly pragmatic necessity, rather than for the explicit and active attempt at documentation. That these representations *become* documentations (become, in archives, the material of theatre history) is a secondary function, if also the default result of the transient nature of live performance. It is for all these reasons that, while drawing in commentary from a wide range of sources, this and the following chapter will (perhaps surprisingly and unfashionably) concentrate on the newspaper review as the archetypal form of written performance representation.

The interest here, therefore, is in the material, immediate, slightly grubby and compromised world of the newspaper review. The intention is to focus in detail on the nature of the review as a medium, looking at its narratives of production and consumption and examining its evaluative, interpretative and descriptive content and functions. Following on from this, Chapter 10 presents more speculative and poetic understandings of how the language of reviews might provide us with presentational knowledge of the phenomenological experience of live performance.

India Song

This chapter, therefore, sets off by explicitly asking what a review is, considering its form, status and purpose and asking how these affect the language employed and the nature of the resulting representation. Loosely grouped around the exploration of narratives of production (and the form of reviews) and narratives of consumption (and the function of reviews), this chapter draws on prominent articulations of reviewing, particularly from critics themselves, from 1945 to today.

However, when providing actual examples of reviewing in practice, the focus is narrower, not asking what the review is for all time and for all people, but what the review is here and now. Indeed, as a tool of analysis this chapter focuses on what the review is, and how it writes its subject in language, in relation to just one particular instance. Throughout, detailed analysis and illustration is in relation to one production: *India Song*, written by Marguerite Duras, directed by Ivo van Hove and performed by Het Zuidelijk Toneel, in the Netherlands in 1998 and at the 1999 Edinburgh International Festival. (This also has the benefit of referring backwards to the discussion of photographs of *India Song* in Chapter 8.) As a detailed case exploration, therefore, these chapters examine the English language press coverage of this production, providing a total of eight reviews:

Rupert Christiansen, 'Silence is not Golden', *The Daily Telegraph*, 2 September 1999.
Neil Cooper, '*India Song*', *The Times*, 2 September 1999.

Mark Fisher, 'India Song', The Herald, 1 September 1999.
Alison Freebalm, 'India Song', The Stage, 9 September 1999.
Alastair Macaulay, 'Duras's play dissected to dreadful effect', Financial Times, 2 September 1999.
Joyce McMillan, 'India Song', The Scotsman, 1 September 1999.
Gabe Stewart, 'Sounds, smells and style ... but little substance', Edinburgh Evening News, 1 September 1999.
Sue Wilson, 'Tales of passion, obsession and tragic isolation', The Independent, 4 September 1999.

Of these eight reviews, four are from national British 'quality' newspapers, two from national Scottish papers, one from a fortnightly trade paper and one from a local Edinburgh evening paper. These eight instances are employed here as a sample of the current status of the review in the United Kingdom, providing the evidential material for illustrations and arguments drawing from the intensely focused analysis of their form and content.

Narratives of production

As might be expected, the form of reviews typically derives from their function and narrative of consumption. At times, however, this is reversed, with the appearance and content of reviews determined by the narrative of their production. It is this that will be explored here, primarily through consideration of where and how reviews appear.

There are several distinct characteristics about the circumstances of the production and publication of reviews, which are typified by these eight examples. Firstly, in terms of the timing of their appearance, in almost every instance these *India Song* reviews were published while the production was still open to the public. In three cases the reviews were published the morning after the first performance. Secondly, the reviews are all brief, ranging from 200 to 600 words.[1] Finally, all the reviews are specifically (and largely exclusively) about the production of *India Song*.

Together, these points describe the typical status of the review, which is brief, often written at short notice, with an immediate topical time reference, a single subject and published as a distinct and particular entity. Such circumstances of production inevitably begin to dictate content. Space, time and frame of reference are limited, leaving little room for abstract thought or large schemes of comparison. Instead the focus is on the description, interpretation and evaluation of a single production with little deviation.

The question here, however, is whether the physical form of reviews, and the particular circumstances of their publication, has an identifiable and characteristic effect upon their content, style of writing and representational impact. Many critics believe that they do. Indeed, Michael Billington, theatre critic for *The Guardian*, writes that 'one's role is partly defined by a set of pragmatic circumstances: the paper one writes for, the amount of space, the length of one's deadline' (1994: xii). Of these, it is often the demands imposed by brevity of space that receive most comment. Jonathan Kalb, of the *Village Voice*, suggests that space restrictions impact not just on the language of reviews but also on the opinions expressed:

> Space pressure makes you mean, makes you put things in severer terms that you'd prefer, makes you express opinions backed by insufficient description. (1993: 166)

Similarly, deadline restraints can be harsh – in some instances there are only one or two hours after a performance finishing in which a review can be written. Consequently, if Kalb highlights the limitations of space, Gordon Rogoff, formerly of *The Village Voice*, stresses the pressures of time:

> The overnight deadline demands a quickly engineered response, a punching bag style, swift and unequivocal judgement matched by easy-come adjectives that just as easily avoid ambiguity and resonance. (1985: 133)

Patrice Pavis makes similar observations, particularly considering style, in his consideration of the press coverage of Peter Brook's *Measure for Measure* in Paris in 1978, remarking that 'The articles often conclude with a paradox, an expression of regret, a metaphor or witty punch line' (1982: 104). Illustrations of this are easily located in the *India Song* reviews, including a punch-bag style, easy-come adjectives and overly stringent condemnation:

> The supposedly oriental smells wafting into the auditorium reminded me of nothing more exotic than Johnson's floor polish. (Christiansen, 1999)
>
> [Het Zuidelijk Toneel's previous productions] were pretentious anatomy lessons; the plays themselves were corpses laid out cold upon van Hove's dissecting slab. (Macaulay, 1999)

[The plot] sounded like Ghandi meets The English Patient. (Stewart, 1999)

It would not be too harsh to judge that the language in these examples is lead primarily by linguistic reflex (the over-used metaphor, the flippant retort) rather that by careful thought. Although this practice is not limited to negative comment – the cliché 'impossible to forget' (McMillan, 1999) being an example of similarly formulaic praise – the *India Song* reviews do demonstrate a much stronger shared rhetoric of 'knocking copy'. As Ivor Brown, formerly theatre critic at *The Observer* notes: 'it is easy to approach a play with a sour quip, ready-made joke, or cutting remark but they are no substitute for truly considered writing' (in Fry, 1952: 36).[2] Writing carefully and accurately about something, positively or negatively, is far harder than merely trying to send it up.

It is possible, however, to over-stress the significance of time restrictions on the content and style of reviewing. Indeed, American critic Harold Clurman pointedly remarks that the reviewer 'infrequently has more to say about a play after a week's reflection than he said immediately after the performance' (1994: 511); while Ian Herbert, founder editor of the *Theatre Record*, suggests that there has been no noticeably greater evidence of contemplation or improvement in style as a result of the gradual disappearance of the overnight review over the last two decades (1999: 242). And from an alternative perspective, Arlene Croce, of *The New Yorker*, argues that the short timescale can actually be beneficial. 'For the dance critic it is best to write as soon after the event as possible', she suggests, in order to retain a fresh and vibrant memory and convey the 'real substance of what went on' (1978: ix). Billington also reflects on this debate, recognising the immediacy that Croce describes but concluding that:

> The overnight review can pin down the exhilaration or the outrage of the moment, but it can also lead to flush excitability and often militates against innovation ... things often look different by the sober light of dawn. (1994: xii–xiii)

However, while critics' complaints about the circumstances of their writing are understandable, it is debatable whether they describe effects symptomatic of all reviewing or merely bad reviewing. What seems more likely is that the complaints portray how restrictions in time and space make the task more challenging. Consequently, linguistic short-cuts, clichés and over-harshness are temptations, far easier than producing considered copy written to a deadline and space restrictions. The examples provided from *India Song* are illustrations of the results of such temptations, but are not the inevitable state of reviewing. While time restrictions exist, therefore, they are not an excuse for bad practice, and while space limitations undoubtedly do impose restraints on the writer, they should not and need not determine style. Consequently, although the form of the review is strongly dictated by the circumstances of its production and publication this does not determine the nature of reviewing.

It is also clear that form is the physical reflection of the function of the review. For surely one of the essential qualities of the review is its temporal near-proximity to its subject: the fact that its composition, publication and (initial) reading are close in time to the performance. Certainly the overnight review also has strong journalistic justification, as from an editorial perspective if it happened the previous night, it should be in the next day's paper. After all, as Keith Bruce, arts editor of *The Herald*, puts it 'that is what a daily paper is all about', with critics reporting upon performances much as journalists report upon news stories.[3] This importance of topicality leads directly to the imposition of time restraints, with this temporal immediacy central to both the form and function of reviews. Length similarly manifests this immediate relationship with its readers and the performance. The ability to read a review quickly provides it with a performative element of its own. Indeed, in many ways reviewing is fundamentally about communicating an immediate response. To change these elements would be to change one of the defining functions of the review, and alter how readers use them. What is certainly the case is that reviewing is kind of practice, a craft that is determined by the nature of the medium and the form that it has assumed in society. The content of reviews is therefore the result of an implicit negotiation within this practice, something that draws attention to the process and act of writing itself.

However, it is perceptions as to the function of reviewing that have, or should have, a more significant impact in guiding content and language. In the following analysis the function of reviews is divided into categories of evaluation, interpretation and description.

Evaluation

For many the very definition of reviewing is its evaluative function, although this need not imply any simple 'it is a hit' or 'it is a miss' awarding of marks for merit or demerit. For example, Clive Barnes, former dance writer at *The Times* and *New York Times*, declares 'I think it's merely the expression of [the critic's] own taste that is important' (in Searle, 1974: 5).

The frequency and explicit nature of evaluations in many reviews corroborates the centrality of evaluation to the practice of reviewing. In almost every one of the *India Song* examples it is possible to identify a clear sentence or phrase where the critics state their overall evaluation of the performance:

> Disorientation and amazement go hand in hand during Ivo van Hove's brilliant production of Marguerite Duras' *India Song*. (Freebalm, 1999)
>
> [*India Song*] is so dull and anti-theatrical that it is difficult to keep one's mind on it in the theatre, let alone recall it afterwards. (Macaulay, 1999)

These two extracts, along with many other possible examples, have a number of things in common. Both are direct statements, occurring at or near the beginning of the review and before providing the reader with any other material or information. Both occupy unambiguous positions of opinion – one is a 'rave', the other a complete condemnation – with such communication of evaluation often appearing to be the very pretext of reviewing.

This assertion of evaluation, moreover, is far from abstract, with the reviewer passing topical judgements that have a direct and current impact on the performance in question. While this is something relevant for all the arts, some observers suggest that the connection is especially tense with live performance – indeed John Booth suggests that 'It is in theatre where, within hours, destinies can be shaped by the nature of critical reception' (1991: 28); while Elsom writes that daily/weekly critics 'are not to be valued for their opinions but for their impact upon trade' (1981: 1) – where the close temporal relationship between review and performance can become an immediate factor in the continuation of a production run.

One function of reviews, therefore, is to act as a source of recommendations, used by readers to help decide whether or not to see a production. And while artists may be uneasy about the position of the critic as authoritative and *public* commentator – as Josette Féral comments, theatre artists 'have a hard time accepting the fact that the critic is not only a self-proclaimed assessor of their work but also someone who can accuse them publically' (2000: 309) – they are far from neutral players in this relationship and seek to utilise reviews to increase ticket sales.[4] One consequence of this is the selective transformation of reviews into publicity quotes: something even shorter, even more quasi-authoritative and even more defined by the communication of value judgement. The review as publicity quote, additionally, renders only positive evaluations of interest. For example, from the *India Song* coverage, it would be possible to imagine the following extracted for promotional purposes and front-of-house display boards:

> This astonishing piece of theatre ... the effect is electrifying. (Fisher, 1999)
>
> Strong, beautiful and bold, and impossible to forget. (McMillan, 1999)

Indeed, *The Scotsman* published the last extract the morning after the first night of *India Song* and it is difficult to overestimate its value in selling tickets for the remainder of the run. With critics very aware of this use of their reviews and the potential for their readers to see the performance it is clear that there might well be direct, if complex, links between the style, language and content and the possibility of impacting on performance production and promotion.[5] This is not necessarily because critics explicitly see themselves as part of the publicity process, but instead a result of the 'genre' of reviewing, of which the snappy evaluation is a significant part.

A more elevated interpretation is to see this same evaluative function and practice as part of the writer's duty not only to inform but also to educate the reader, becoming a kind of 'gatekeeper' to the arts. John Booth, for example, suggests that the critic as gatekeeper assumes responsibility for maintaining high standards in the arts (1991: 160); Wesley Shrum also argues that critics, as 'taste makers and gatekeepers', are part of a system that seeks to 'grant knowledgeability a role in the ascription of quality' (1996: 96). Similarly, as an arts editor Bruce sees the role of his newspaper as a kind of gatekeeper, telling its readers what is on and what is interesting. Moreover, he describes the most common reason for changing or rewriting a review as being a result of lack of clarity in the communication of the critic's opinion, with the writer taking a hazy or compromising position on whether they liked a performance or not. For Bruce readers primarily want to know what the critic thought of a production – in terms of evaluation – with this crucial to the nature of reviewing. In this manner the critic becomes arbiter of what is good and what is bad, and governs the initial status of new works in the cultural canon.

The review, therefore, possesses an authority that makes it more that just one audience member's opinion, although unhappy artists may well dismiss them as such. This status derives partly from the reputation of the writer, partly from the newspaper itself, but most significantly is a product of simply appearing in print at all. As Susan Melrose puts it, publication within a newspaper means that the expression of taste 'functions not as anecdote but as *expert discourse*' (1994: 77). This is what Hans Keller, in his condemnation of music criticism, calls 'the black magic of the printed word, which lends authority where there is no authority, interest where there is no interest, power where there is no force' (1987: 191). Illustrating the practical experience of this, Joan Cass, dance critic on the *Boston Herald* and *Dance Observer*, describes how she realised that the appearance of her own words in print conferred the opinions she expressed with an automatic air of authority:

> The printed word has a powerful authority. I have found myself reading dance reviews in the morning newspaper, with respectful attention, despite the fact that I, the writer, knew I left the concert hall the previous evening uncertain of my opinions. (1970: 225)

Despite all the limitations and possible shortcomings behind their production, therefore, once in print the review obtains, for better or worse, a semblance of stability, firmness of opinion and authority in evaluation. This performance of power that is enacted by the process of criticism once again positions reviewing as a form of writing that is also a form of doing, a form of continuing action.

Interpretation

The critic, therefore, operates somewhere between the artist and performance and the (actual, potential or theoretical) audience. This is a position that Clive Barnes describes in terms of responsibilities in both directions:

> I think a critic is trying to build a bridge between the two, that he's trying to help the artist by helping the public understand him. (In Searle, 1974: 7)

Barnes suggests here that reviewers, positioned between the artist and the audience, have a role in aiding communication and understanding between the two. There is a subtle distinction here between the description of critics as 'bridge' and as 'gatekeeper' or 'taste maker'. Although both maintain the idea of the reviewer as speaking for the art, there is a change of emphasis from evaluation to interpretation: from judging to bridging, explaining and elucidating.

The interpretative mode most frequently presented in the *India Song* reviews is one that seeks to explain to the reader the intentions of the director or playwright, with the critic first detailing the theatrical technique used and then suggesting its significance. For instance, several reviews discuss the fact that the play is staged with the dialogue (and stage directions) pre-recorded and relayed over loudspeakers:

> In the script, no word is spoken in view of the audience, the action being described by offstage voices, a technique used to emphasise the onstage torpor. (Fisher, 1999)
>
> The separation, for the most part, of the actors from the dialogue seems intended to imply their powerlessness in the face of larger, destructive forces. (Wilson, 1999)

In these examples, the critics seek to help the reader understand what they see as the artist's intention. Illustrating the thinness of the line between interpretation and evaluation, the writers then typically also comment on whether they consider the technique to be successful or not. This appears to be based on the age-old critical assumption that 'good' art is that in which artists fulfil their intention, something that requires reviewers to first discover the intention and then assess how successfully it has been realised. Noting this critical reflex in her own work, dance critic Deborah Jowitt writes self-reflectively in a review of the Nederlands Dans Theatre:

> [Choreographer Glen] Tetley works very skilfully at preparing the audience for the first live naked body, so skilfully he almost achieves the reverse of what he intended. (Who knows what he intended? Typical shoddy critic-talk...) (1977: 40)

Rarely, however, are reviewers as self-reflective as this. Indeed, interpretation is generally only present at all where the writer has a positive response to the production. Where opinion is negative, the impulse to interpret the artist for the reader is frequently replaced by the criticism of the intention and its realisation. Again the following example centres on how the pre-recorded dialogue was divorced from the live actors, this time however, there is little attempt to explain or analyse the technique except in terms of evaluation:

> Its novelty is that the actors on stage, barring two climactic moments, do not speak: they only mime the action as the tale is chorically

related by unseen narrative voices. The device is not particularly fruitful: it becomes confusing, and it led the actors here to some ludicrous silent-movie excesses. (Christiansen, 1999)

Here, the critic does refer directly to questions of performance: the associations created by 'mime' and 'silent-movie' constructing the beginnings of a description of the performance style. However, he immediately imposes judgement on these elements: while the evaluative declaration ('ludicrous', 'excesses') is extremely clear, the descriptive content is much more ambiguous. Any interpretative content, meanwhile, is obscured.

In the eight *India Song* reviews as well as interpreting stage effects, the critics also suggest overall interpretations. They offer readings as to what the playwright intended, what the characters are like, what their emotions represent, what allegories are present and what the play as a whole means or symbolises. At times this can be at a very sophisticated level, such as with Joyce McMillan's review in *The Scotsman*, where she discusses how director van Hove explores Duras' central concern with 'the confrontation between a Western culture based on "masculine" ideas about action, control, order, and an Indian culture based on passivity, fatalism, acceptance of human life as part of a huge organic cycle of decay and rebirth' (1999). This is perhaps the only instance where the critic is primarily (or even significantly) interpretative in outlook.

Generally, however, even when not based on intentionalism, it is often impossible to tell whether critics make interpretations on the grounds of taste, technique, tradition or form. Indeed, it is often impossible to distinguish between interpretative and evaluative modes, with one recurring element in the *India Song* reviews being the opacity of the opinions expressed. The result is that all too often it is impossible for readers to work through the reviewers' interpretative positions for themselves, or, alternatively, reach a different understanding of their own from the information given. This is what Kalb describes as 'insufficient description', leaving the reader unable to test or otherwise evaluate the interpretative conclusions expressed so firmly by the reviewer.

There are many possible explanations for this, including time-and-space-induced pressures to deliver snap judgements backed by insufficient or non-existent description. Additionally, the articulation of evaluative judgement *is* central to the nature and function of reviewing as a form; the practice of reviewing is the act of asserting opinion. More crucially, however, it is the framing of interpretation and analysis through the deduced or presumed intention of the artist that produces difficulty. Possibly as a result of seeking to bridge the divide between artist and audience, this focus on intentionality means that critics seek to speak for the artist rather than of the work, and rather than letting readers see for themselves through the eyes of the writer as audience member.[6]

Description

Previous chapters have explored the motivation prompting the creation of archives, photographs, video and other enduring documentations of live performance. In their different ways, all such post-performance existences are representations of performance that enable some trace of the event to remain after the transient moment of its production. Reviews, and writing more generally, are equally motivated by an impulse to 'record' and thereby retain something of an ephemeral performance. Deborah Jowitt, for example, describes how her writing is stimulated by an 'anxiety to capture and chronicle a notoriously ephemeral art' (in Copeland, 1998: 100); similarly, Marcia Siegel describes one responsibility of the critic as that of 'reporter', who enables 'dance to have a history' (1977: xv). Many other writers repeat such ideas in one form or another, such as Shrum who writes that the ephemerality of performance 'is what gives reviews their special power and significance' (1996: 12). Indeed, even if not consciously or explicitly talking about the need for documentation, many critics articulate the review as a forum through which performance is provided with an enduring history.[7] Writing lasts in a way that live performance does not, so that for Kenneth Tynan:

> what turned me into a critic was the urge to commemorate those astonishing men and women whose work would otherwise die with the memories of those who saw it. (1984: 12)

There is also a degree of consensus as to how to achieve this ambition: a writer 'records' a performance through description. Of course, description is not a direct, mechanical or authoritative method of recording in the manner of still photography or video – although appearance in print and the status of the critic as 'expert' do provide reviews with a kind of authority. Description, however, does not literally document or mimetically record the performance. Nonetheless, the invitation to the reader to imagine the performance through description certainly is an attempt at representation, at reporting experience and recording the event.

In contrast to a review offering primarily evaluative or interpretative comment, considered description, it is argued, provides readers with detailed information about the performance. Further, description allows readers to look beyond the critic's opinion and obtain some knowledge of the appearance and nature of the performance in its own right. As New York critic Frank Rich puts it:

> For me, passing judgment on a play is absolutely the least interesting part of the job ... The creative part of the job, the reason I enjoy doing it, is to try to re-create for the reader the experience of what it was like to be in the theater and see a particular play. (In Booth, 1991: 176)

This is an ambition echoed many times by other writers: such as Edwin Denby, who states that 'what one enjoys most in reading is the illusion of being present at a performance' (1986: 539). Similar formulations are also used to praise the work of particular writers, with Julie Van Camp celebrating the work of Arlene Croce for how she 'uses words to capture a sense of what it was like to be in the audience' (1992: 42); likewise, Robert Brustein praises Tynan for always managing to establish the 'exact verbal equivalent of the visual events he had witnessed' (in Booth, 1991: 178).[8]

Although these calls for a descriptive approach to reviewing are often urgently expressed in relation to the live arts, the most influential exponent of a descriptive criticism was not writing about performance at all. Instead, Susan Sontag's *Against Interpretation* is a powerful exposition of the need for a descriptive critical writing about literature, and less particularly all arts. This is presented as a moral endeavour, with Sontag taking a judgemental position against what she regards as the dominant instinct of the critic, interpretation: 'To interpret is to impoverish, to deplete the world – in order to set up a shadow world of 'meanings' (1967: 7).

Sontag suggests that interpretation is secondary to the experience of a work of art, not enhancing the original encounter but undermining its immediacy and destroying its power. Indeed, for Sontag the central and damning argument is that while interpretation does enable communication about a work of art (which she recognises as necessary) this is only because it tames the experience and makes art comfortable and manageable. She suggests that the very striving for interpretation expresses a lack of ability to respond to the experience itself and to what is really there. What matters instead is the primary 'pure, sensuous immediacy' of the experience and Sontag demands a criticism that consists of 'accurate, sharp [and] loving description' (1967: 12).

Sontag's ambition is seductive, appearing to offer hope for the admirable goal of representing the experiential perception and value of art. Moreover, her image of the skilled describer of the arts has, when appropriated by performance writers, provided one of the most influential manifestos for the form. This has particularly been the case in dance and performance art, with Sally Banes recording how for her and many other dance writers *Against Interpretation* was a 'sacred text for my generation' (1994: 7). Sontag's essay, Banes argues, was one of the main influences in establishing an aggressively descriptive, anti-interpretative, philosophy that dominated dance reviewing, particularly in the 1960s and 1970s (as can be seen in the opinions already presented from dance critics such as Croce, Jowitt and Siegel). Jowitt acknowledges this herself, saying in an interview that 'Some of us were very influenced by Susan Sontag's *Against Interpretation*. We wanted to confront the work without considering anything but its sensuous apparition' (in Michelson, 2002).

Away from dance, another exponent of descriptive criticism writing at a similar period to Sontag is Michael Kirby, who in an essay titled 'Criticism: Four Faults' rages against what he terms the immorality of evaluation and uselessness of interpretation in reviewing. To replace such 'primitive and naïve, arrogant and immoral' criticism, Kirby calls for a discipline of 'performance analysis'. He sees this as enacting the recording of ephemeral events by description and analysis, avoiding as far as possible conscious subjective statements and even any words that may be interpreted subjectively (1974a: 66).

These advocates of descriptive criticism all suggest that one major responsibility of the writer is to represent or record the performance, thereby allowing readers to know the absent performance through the mediation and language of the review. Given the non-mimetic nature of language, however, the extent to which it is possible for description to 'record' and allow readers to 'see' the performance is certainly debatable. Nonetheless, this passionately expressed ideal is a useful tool against which to test the descriptive content of theatre reviewing today, particularly as represented by the *India Song* reviews.

Describing *India Song*

The *India Song* reviews certainly demonstrate Kirby's perception that evaluation and (to a lesser extent) interpretation dominate much criticism.

Nonetheless, the reviews do contain a degree of descriptive content, including physical description of the stage appearance, narration of the plot, reports on the actors' performances and discussion of the direction and other stage effects. The reviews also include description of the effect (on the reviewer and/or the audience) of any of the above elements. In almost all such instances, however, description slips directly into interpretation or evaluation. To take an extended extract as an opening example:

> [*India Song*] is a strange, elegiac story of doomed love and obsession between the French ambassador's wife – the beautiful Anne-Marie Stretter – and a vice-consul from Lahore, set among the embassies and residences of Calcutta in the dying years of the empire. (McMillan, 1999)

The emphasis here is firstly on the plot and its expressiveness, with these sentences indistinguishable in many ways from a hypothetical review of the play-text of *India Song* or of a novel. In the review this is then linked to the form that this narrative takes in performance, 'Van Hove's production makes the whole audience, drawn into the circle of thick yellow light and sound, part of that culture of decadent voyeurism and gossip' (McMillan, 1999). This connects plot to staging and takes the description away from consideration of something purely literary. This almost natural, instinctive movement immediately begins to communicate something of the performative medium, intrinsically binding content and form together.

In comparison to this narration of plot and staging, however, there is surprisingly little attempt to present direct and detailed descriptions of the performers – in terms of appearance, action or movement – which might be considered an area where the performative nature of theatre might most come across in writing. It is possible to see plot as more familiar and accessible as a textual element and not distinctive of the performative. Similarly, static stage objects do not invite description in dynamic language. However, kinetics, movement in space and non-verbal interaction between humans are vital components of actors' performances, requiring representation in dynamic description. Such description could employ, usefully and naturally, a vocabulary that represented the live performance as grounded in human bodies and human-scaled space.

Examples of this include occasions when the physical nature of the performance is located in phrases such as the 'characters mouth to half-remembered dialogue' (Freebalm, 1999) where there is a kernel of an embodied writing. Another review also pays attention to the stage performances, first contrasting the vice-consul's 'obsessive frenzy' with the way Anne-Marie Stretter 'moves passively around the men'. Later the reviewer comments more directly on the performances:

> Steven van Watermeulen's louche, chain-smoking Michael Richardson ... is a figure of buttoned-up restraint who only comes alive through Anne-Marie. (Cooper, 1999)

Here, in the phrase 'a figure of buttoned-up restraint', it is possible to see something of an evocative, bodily language that draws the reader into the experience by proxy. This phrase uses physical appearance as a metonym for wider characteristics: 'buttoned-up' extends outwards from the bodily experience of a tight top button to convey the experience of constriction, suffocation (literal and metaphorical), uprightness, correctness and stiffness. There is a movement here between literal description and figurative analogy. It is a phrase, finally, which uses the body as its descriptive focal point, a possibility that is developed in detail in Chapter 10. However, such readings probably place an unstable amount of emphasis on single, isolated phrases and overall there are remarkably few examples in the *India Song* reviews of discussion of the performers at all. When they do so, additionally, it is in largely evaluative, non-dynamic statements. The prominent and important evaluative function of the review partly necessitates such assessments, yet such language makes little attempt to enable the absent reader to reach an understanding of how the performance appeared in front of the audience.

Without strong descriptive content, the *India Song* reviews frequently demonstrate a fairly established progression in their content: description of plot, followed by what the director has done to it, followed by what it 'means', invariably followed by a judgement as to its degree of success. This rapid shifting of description into interpretation and evaluation is primarily a result of the critical imperative of reviewing as a form. In this evaluative (sometimes even manipulative) bias, the *India Song* coverage is probably typical of newspaper criticism as it is practised today.

The function of criticism

These problems recognised, it is necessary to consider what kind of description is possible, although it is no doubt easier to establish what kind is impossible: including exhaustive, correct or objective description. Indeed, the foremost response to Kirby's demand for a neutral criticism – free from personal or pseudo-objective judgement – must be to seriously question whether any description can be wholly objective. That Kirby insists that words such as 'beautiful' are primarily pseudo-objective, rather than descriptive, suggests the difficulty of a descriptive writing that is not evaluative. Writing cannot directly document or transcribe performance, instead by necessity description represents the performance as seen and understood by the critic, with choices in language reflecting personal perspectives – and honestly done this is all that can be asked for. This said, however, there are clearly degrees of engaged and communicative good practice in descriptive reviewing and of what might legitimately be labelled uninformative and imprecise bad practice. Good practice brings the performance closer to the reader; bad practice distances it behind evaluation, interpretation, rhetoric and wit – all of which have their place, provided it is not one that occludes the representational and descriptive.

Kirby himself provides no examples of good technique in his discussion of bad practice. Similarly, while Sontag is specific about what she thinks criticism should do, she is less explicit on what language and techniques it should employ. While she suggests 'clear, sharp, precise' description, she never illustrates what this is or how to do it, all the while admitting that it is very difficult. Instead, she merely lists examples of 'good' descriptive criticism (Manny Farber on film, Dorothy Van Ghent on Dickens, and Randall Jarrell on Whitman) without providing detailed examination of what makes them good. Additionally, closer inspection of one of these examples is interestingly unhelpful. In 'Some Lines from Whitman', which Sontag praises, Randall Jarrell directs his attention to how Walt Whitman writes, rather than what he writes about or how he can be interpreted. In this, Jarrell certainly practises Sontag's prescription for a non-interpretative criticism that considers the reader's experience of the work; but his method of performing this intention is to quote Whitman directly, at length and with little addition. At various points Jarrell remarks upon this practice:

> To show Whitman for what he is one does not need to praise or explain or argue, one needs simply to quote ... How can one quote enough? (1960: 106–110)

Finally, Jarrell notes that, as with much great writing, Whitman's work achieves 'a point at which criticism seems not only unnecessary but absurd' (1960: 119). The implications of this statement are extremely problematic.

It is possible that quotation performs 'criticism' through demonstration. Surely, however, it only performs descriptive criticism if the idea of description is exact replication of the original – the logical result of which would be to replace literary criticism with reprints. That literary criticism is manifest in the same medium as its subject (words about words, as Hans Keller puts it), makes this possible if not desirable. Similarly art criticism might be replaced by reproductions, exactly what George Steiner proposes in *Real Presences* when he argues that 'dispassionate summaries', 'representative extracts and quotations', 'catalogues' and 'reproductions' are the only discourses about art that are necessary or legitimate. Equally, Steiner suggests that all performance criticism is parasitical and secondary and that in an imaginary utopia of 'immediate responses' it would be replaced entirely by repeat performances (1989: 5–8). Aside from its practicality, the desirability of this utopia is arguable. First, however, as a method of teasing out exactly what the function of performance criticism is (or should be) such proposals are worth pursuing further.

The implication of Steiner's condemnation of secondary responses to art would seem to suggest that the only valid commentary would represent its subject as directly and untransformingly as possible. Echoing such ambitions, Patrice Pavis appears to lament the fact that 'no description can do other than radically modify the object it describes' as if the objective should be perfect reproduction (1982: 111). This also seems to be implied by Jarrell in his extensive use of quotations: in using the very words of his subject he appears determined to remove all risks of modification. Perhaps the implication of any assertively 'descriptive' criticism, such as that advocated by Sontag, would therefore be writing that, if it did not literally reproduce its subject, would aim to reproduce in the reader the experience of its subject.

An example of this problematic desire for criticism that replicates the experience of its subject is put forward by Roland Barthes in 'The Grain of the Voice' (1985). In this essay, Barthes argues that music 'fares badly' (267) from the onslaught of linguistic translation, a remark similar to Steiner's declaration that 'When it speaks of music, language is lame' (1989: 19). Music criticism, Barthes suggests, produces writing that is about writing about music: writing that is one, possibly two steps away from the music itself. Barthes is largely right here: music reviewers frequently devote large amounts of their limited space to contextual aspects – including performance history, social history and biographical and psychological interpretations – with any consideration of the actual performance sidelined as a result. (Such deviation, additionally, is not restricted to music and is noticeable in much other reviewing, including the *India Song* coverage.) While examination of these issues is neither uninteresting nor unimportant, Barthes is correct that their consideration has a tendency to dominate over description and representation of the experience and of the performance. This tendency is no doubt encouraged by such extra-musical discussion having a very strong, shared vocabulary and established discourse, in contrast to the weaker discourse of performance representation. (And no doubt also by the fact that it is much easier.[9])

Indeed, while Barthes criticises writing about music for having a strong 'pheno-text' (all that is literary, meaningful, analogical, contextual), he complains that it has a weak or non-existent 'geno-text' (the practice, the action, the experience of music, ending not in understanding but in pleasure or bliss). Barthes' separation of pleasure from understanding mirrors Sontag's anti-interpretative and pro-experiential agenda and the privileging of sensual experiences over intellectual responses that Banes (1994) and Copeland (1998: 104) argue is the motivation for descriptively biased dance criticism. Although this division is artificial – in my experience pleasure and understanding often co-exist – it is apparent that much contemporary music and theatre criticism (and to a lesser extent dance criticism) does include a predominance of phenotextual analysis. To perform 'well', therefore, Barthes argues that critical writing must somehow develop its geno-text. The ambition for Barthes is for language to account for music while losing or modifying as little as possible of the nature of the original experience (the similarities here with Sontag's 'erotics' of art are evident). He is looking for a language that can somehow replicate the experience of music: music not as meaning but as practice. In this, Barthes maintains his demand that criticism must always write in a language harmonious with that of the work (Davidson, 1968: 98).

The logical extension of Barthes' ambitions, and his suggestion that language performs badly in accounting for music, would seem to return to Steiner's argument that only the performance of music presents useful criticism about music. As it is self-evidently impossible for language to quote music directly, perhaps the only method of fulfilling Barthes' ambitions would be musical analysis through music. Similarly, perhaps the only useful communication about performance can be through performance, suggesting by implication that all other extra-performance discourses must ultimately fail to represent anything approaching the experience of their subject. Putting such theory into practice, Hans Keller proposes a system of 'functional analysis' that seeks to avoid all the transformative, evaluative and distorting effects of language about music. Overcoming the obstacle of criticism being in a different medium and 'language' than its subject, Keller's method of criticism would be wordless, 'notes about notes, as literary criticism is words about words' (1994: 8).

Keller's concern is with the demands of the formal analysis of musical compositions, not with music in performance. However, it is interesting that he expresses particular delight in the performance of his 'functional analysis', and satisfaction at the positive audience responses to their live performance (1987: 147). That Keller saw the fulfilment and justification of his ideas in their performance is one reason to be sceptical as to whether such proposals present actual, practical or useful methods of music criticism. At one point Keller declares that his analytic scores are easier to understand than the originals because they bring the background of their subject score to the fore of the functional analysis (1994: 127). In which case why go to the bother of listening to the original? All that is important would appear to be in the analysis, and, additionally, in a more accessible and filtered form.[10] Although no doubt Keller does not consciously intend this, if a functional analysis represents what is essential about a piece of music then, by implication, everything else is inessential. Hence, functional analysis would either tend towards replication of its subject or render the original logically obsolete. Musicologist David Burrows makes a similar point when he discusses possible methods of presenting (as a form of criticism) experiential responses to music. Burrows suggests that any such method

> would have to provide an experiential inventory of every detail in it, and in the order in which they occur ... But besides the tedium entailed, this procedure would be too close to the actual musical experience to accomplish certain independent aims of analysis. (1997: 540)

An experiential analysis, therefore, might be too close to the actual experience of music to be useful as analysis: an effective retort to the demands for a criticism that somehow replicates the semiotic system or sensual experience of its subject. Such criticism fails theoretically and practically because it is too close to its subject. As Josette Féral writes, if a critic limits their intention in seeking to transparently 'echo artworks in an enlightened fashion' then this is both 'necessarily insufficient' and to neglect the real function of criticism. Instead, for Féral it is necessary for all criticism to be translations, both into the different medium of words and into the critic's vision of the art – and if this is also enacts a form of betrayal of the art then that is also necessarily (2000: 309–10).

The function of language-based criticism and commentary about art is to enable communication about art, providing it with a presence in the exchange and circulation of meanings and experiences that is our cultural world. To do this some kind of distance, and indeed translation and modification, is both required and valuable. The ambition of criticism, therefore, is not to be neutral and complete, nor to replicate its subject. Instead, it is precisely to be selective, presenting what the writer finds interesting, memorable and worth articulating about the experience.

The radical and ideologically driven perceptions of the parasitical and destructive relationship between art and language appear to ignore the nature and interest of discourses about art. In particular, this neglects the importance of language in enabling art, as a social experience, to become shared in subsequent discourses. Discourses about art are fundamental to the very experience and nature of art, the result of a sense that the experience needs to be interpreted, discussed and re-communicated in order to be fully completed. Additionally, while language evidently transforms primary responses, the attempt to articulate experience in language has great value in revealing individual and social perceptions about art and artistic experiences. Discussion about the experience of performance – imperfect, mediated by language and reflecting either conscious choice of words or employment of a culturally established and shared vocabulary – provides a window into cultural perceptions of the thing itself. What is required, therefore, is not the rendering of a perfect reproduction of the original performance into language (which is both impossible and undesirable). Instead, Chapter 10 explores whether it is possible to produce description that evocatively renders the critic's evaluative, interpretative and sensual experience of the live performance in writing.

10
Writing the Live

Much of the discussion in the Chapter 9 explored typical practice in daily reviewing, including examples of what might unashamedly be labelled typical bad practice. This chapter, by contrast, pursues more speculative possibilities of good practice, identifying and describing instances where writing does begin to represent something of the experience of live performance.

The previous chapter explored calls, from commentators including Barthes, Sontag and Kirby, for an expressive, evocative and experience-centred writing about art and performance. At the same time it explored how criticism should and could not aim to record a performance in any objective, complete or comprehensive manner. Instead, I argued for a form of writing that openly represented the writer's subjective experience. Such representation would be partial, transformative, evaluative and interpretative, yet crucially would also focus on the lived experience, on the distinct character of being there in the auditorium and on the particular nature of the performance as theatre, as dance and as live performance.

This endeavour bears similarities to other explorations into the possibilities of performative writing, particularly Della Pollock's 'Performing Writing', which advocates the evocative, active and creative possibilities of critical writing:

> Performative writing evokes worlds that are other-wise intangible, unlocatable: worlds of memory, pleasure, sensation, imagination, affect, and in-sight. ... It does not describe in a narrowly reportorial sense, an objectively verifiable event or process but uses language like paint to create what is self-evidently a version of what was, what is and/or what might be. (1998: 80)

Pollock's interest in 'making writing perform' (75) and in what writing does in the 'interplay of reader and writing in the joint production of meaning' (80) is key to the interests and intention of this chapter. Unlike Pollock, however, this chapter is also interested in another kind of practice, with again the focus largely on the medium and practice of newspaper reviews. Once more core illustrations will be drawn from the eight *India Song* reviews examined in the previous chapter, but now also accompanied by an expanded range of examples including the writings of significant performance critics of the twentieth and twenty-first centuries – Michael Billington, Edwin Denby, Andrew Porter, Kenneth Tynan and Marcia Siegel. This maintained focus on reviewing is partly directed by its position as the most widely produced and consumed form of written performance documentation. Additionally, reviews are a medium that is very much about the activity (the doing) of reviewing, with meaning constructed within the unfolding and performing of this practice. For both these reasons reviews are in some sense the archetypal form of written performance representation and a medium that it is valuable and necessary to engage with, interrogate and to challenge. And in challenging reviewing practice, this chapter develops and imagines the potential of written representations of live performance, engaging with concepts of phenomenology and testing the possibility of encoding liveness into the very language and structure of writing.

The ambition of engaging the reader with a sense of the critic's own experience of a performance is something that, as explored in Chapter 9, is often presented as the benchmark of good criticism. As Eric Bentley puts it: 'Nothing a critic has can open your eyes except his own eye: he says, look! And you look ... A good critic will get you to do so' (in Booth, 1991: 176). The first step in doing this, it is suggested, is a detailed and precise description of the performance. If the eight *India Song* reviews examined in Chapter 9 are taken as typical, however, then most reviews include little significant description whatsoever, let alone what might be identifiable as an evocative description of the lived experience. Moreover, it is necessary to differentiate between straight description (using language as referent to what is absent) and 'generative and ludic' writing that positions the reader as encountering the language itself (Pollock, 1998: 80). In terms of the description of liveness, language as referent might be seen in terms of factual descriptions, perhaps noting something distinct about any particular performance, such as with this fairly characteristic example:

> Composer Harry de Wit's presence on Jan Versweyveld's set playing his score live gives the production the air of a precise avant-garde

concert, and one is gripped by the actors' unstudied concentration. (Cooper, 1999)

This kind of occasional explicit reference, however, is of less interest than any implicit aspects within the very structure and grammar of a review that might signal a more unconscious, embedded and generative perception of the event as a live performance. In this extract, for example, the words 'live' and 'presence' are factual rather than evocative. In contrast, when the critic writes that 'one is gripped' it is possible to see this as the beginnings of the construction of a present, dynamic and continuing event. It is this kind of subtle, perhaps unconscious and certainly implicit evocation of liveness that this chapter focuses upon. The interest is in how an unspoken conception of the live manifests itself in the language of reviews; moreover, how would it be possible to communicate the temporally located and spatially embodied experience of the live as an intrinsic part of the very structure and language of our writing.

To begin to mark out where such possibilities for an experience-centred right might be located, it is worth starting by considering short segments from the different *India Song* reviews, each presenting the same element of staging. The following extracts all discuss the production's attempt to create a sub-continental atmosphere, particularly through the use of 'smellorama' effects.[1] Between them, they display the whole range of descriptive methods and habits explored in the previous chapter. The following analysis briefly discusses each example in turn, starting with instances of 'bad' descriptive practices (the immediately evaluative, manipulative, pseudo-objective, overly rhetorical) before highlighting how some contain within them the possibility of a more dynamic and evocative language. The first example is unambiguously negative:

> the staging failed to conjure an atmosphere of torrid monsoon dampness, and the supposedly oriental smells wafting into the auditorium reminded me of nothing more than Johnson's floor polish. (Christiansen, 1999)

Here even before the writer employs the descriptive element 'oriental', he already qualifies and evaluates it with the very judgmental 'supposedly'. It may be that the effects were not, in the critic's opinion, effective, but the review continues with an entirely pithy comment drawn from an established rhetoric of 'knocking copy'. The second example is another instance where the primary motivation seems to be to display the reviewer's own wit:

> An aroma-stimulating technique sounded promising, but instead of street smells, spice or the scent of the monsoon, the overwhelming aroma was of lavatory cleaner. (Stewart, 1999)

This, in fact, is not a description at all, but an illustration of manipulative evaluation designed to bring the reader into agreement with the critic's opinion while bypassing the actual production and original experience. The implication of 'sounded promising' is the immediate supplementary 'but instead', meaning that even before proceeding any further the reader is forced into grammatical concurrence with the critic.

Other reviews, however, begin to draw the reader closer into the writer's temporally located and physically embodied experience of the performance. For example, one review describes the performance in terms of the audience experience of the stage design, describing how the audience should

> go prepared to loll – literally, if you have a seat on the stage – on large black silk cushions, drenched in hot light and sudden inky darkness, bathed in unexpected and intense sounds and smells, experiencing something unsettling and strange. (McMillan, 1999)

Here words such as 'loll', 'bathed', 'drenched' and 'unsettling' detail the experience and also the bodily nature of and response to the experience. Such words locate the audience in the theatre as present physical beings and immediately begin to create an experiential impression of the performance. Although the actual, identifiable description of smell in this passage is minor and non-specific its function is more significant, rooting the description in time and place and powerfully working on the reader's perceptions. Here smell is placed in what might be described as a synaesthetic position, combining with colour, texture, sounds and emotion to form a multisensual picture of the performance in the mind of the reader. While several words, such as 'unexpected', 'intense' and 'unsettling' hover uncertainly between description and interpretation, as subjective statements they tell of the writer's perception of the performance.

This review also evokes a sense of the lived experience of the event, with a very physical description of the audience that provides a bodily (or communal and multi-bodily) focus that grounds perception as humanly located and active. In ensuring that the reader is aware that the performance is an event experienced by people, the review resists perceptions made in abstraction and additionally allows the reader to imaginatively enter into that experience by proxy. Another example also seeks to evoke the atmosphere for the reader:

> washes of citronella engulf the audience, completing a sensory bombardment that sets the nerve-endings aflame. Never can the sight of six silent actors drifting across a stage have been so riveting. (Fisher, 1999)

The evaluative element of this extract is again most readily apparent, 'never' really stating nothing except that the critic thought it good. However, what is also interesting is the use of a large amount of dynamic imagery in the present tense – 'washes', 'engulf', 'bombardment', 'aflame', 'drifting' and 'riveting', all words which contain at least a degree of synaesthetic connotation – inviting the reader to create a picture of not so much of what the performance looked like but what it felt like to the writer. These kinds of words describe the performance in a manner that also details the witness' emotional or mental response to the experience, often communicating an emotional response that it is possible to reflect in embodied posture, attitude and feeling.

Indeed, the language is more evocative than descriptive, using both sight and smell perceptions to communicate the entire bodily experience of being there. This is an experience where, again, the viewer is actively present: with the act of seeing, of being audience to the performance, manifest in the language. In doing so it contains the beginnings of an effective and dynamic description that invites the reader to access (through creation not recreation, representation not reproduction) the critic's experience of the live performance.

These identifications of an evocative and experiential writing of performance are tentative. Although occasionally present it is not central to the language of the reviews, which instead of seeking to represent the nature of the experience focus much more strongly on evaluation, interpretation and contextualisation. Nonetheless, particularly reading across all eight reviews of *India Song*, it is possible for readers to begin to imagine themselves into the experience of the production through identifying these linguistic traces of a momentary and embodied experience. What this also does is to point us towards possible areas where we might be able to develop a true writing of the live experience of performance and each of the possibilities fleetingly identified here – for the writing of performative time, the evocation of a sense of space, of embodiment and the use of dynamic and synaesthetic language – will be explored in turn and in greater detail in the following analysis.

Writing time

In attempts to articulate the ontology of live performance, imagery of time is prominent. Words suggestive of time – such as trace, disappears, memory, moment, event and present – recur repeatedly, as performance is described as passing in front of an audience, as an event for that moment only, as located in an ever changing now. When Peggy Phelan suggests that performance becomes itself through 'disappearance', or Michael Kirby insists theatre is 'ephemeral', or Merce Cunningham declares that dance is fleeting, each is evoking the determined and limited temporal location of the live event. The concept and ideological valuation of disappearance is itself a description of how performance exists, or rather passes, in time. Equally, ideas of time are central within discourses of documentation, whether evoked in the photographer's ability to freeze and capture a moment of a performance or the archive's status as a record of time past and performance history.

However, alongside the host of theorists and practitioners who explicitly use time-based imagery to describe the nature of live performance, few have directly examined how writing might grasp this temporality in language. Two writers who have done so are David Burrows and Bernard Beckerman, both of whom follow a direct path from articulating an ontological description of performative temporality to asserting a practical need and desire for this to be communicated in the substance of critical language. Beckerman's starting position is that theatre is a uniquely temporal art form, his objective to establish how performance commentary can best cope with this dynamic aspect.

In doing this Beckerman first describes how conventional theatre criticism adopts what he calls a 'horizontal approach' to analysis, following through entire strands independently: plot, character, spectacle and theme are all disentangled from one another and examined separately from the beginning to end of the performance. This does indeed appear to be the conventional approach of many theatre reviews, with critics frequently structuring their writing through separate consideration of different elements of a production, often in neatly divided paragraphs. This clean (if somewhat wooden) segmenting benefits critics in easing their task compositionally, and does provide an obvious structural guide to the reader.[2] However, as Beckerman observes, it does militate against consideration of the production as a coherent whole, and (in particular) prevents the reader from gaining a sense of the temporal movement of the performance.[3]

Beckerman's alternative proposal is for a 'vertical method of analysis' that examines the same aspects of plot, character and spectacle but now as they relate to coherent sections of the play:

> The vertical method ... is bound by temporal progression. Rather than treat a play as movement along separate paths of plot, character, and so forth, it envisions the progression of the play in its entirety. The elements of analysis, therefore, are not plot and character but units of time. (1979: 37)

Beckerman, unfortunately, does not provide detailed examples of his system in action, nor suggest instances of what he considers good practice. Additionally, the systematic nature of Beckerman's approach is more suited, not least in terms of length, to scholarly considerations of performance and does not necessarily complement the form and function of the newspaper review. Nonetheless, aspects of the approach are worth remembering, particularly for the manner in which it attunes to important articulations of the performance experience as rooted in the live moment.

Like Beckerman, Burrows' point of departure is a discussion as to the temporal nature of performance, this time of music, writing that:

> the coupling of the flow of sounds with the attention of perceivers is controlled by the temporality of the sounds, and is therefore limited to a now whose content changes ceaselessly. Music takes place in its own almost total sonic absence. (1997: 529)

It is worth stressing the similarities, in assumptions, in tone, and in language used between this description of music to that of theatre and dance. Particularly, in this instance, the use of key dynamic words, with 'flow', 'change' and 'absence' standing in here for Phelan's idea of 'disappearance'. 'Sonic absence', for example, suggests that even as music is it disappears, and that this governs the listener's experience.

Burrows' next step is to call for a method of critical analysis that accounts for this temporality, something which he terms 'processual analysis' and that would focus on the 'performance of music', the unfolding dynamic experience, and not on the 'piece' of music as a unitary, static and stable entity. Here Beckerman's 'sequence of total experiences' can be directly compared to the experiential account of music Burrows is advocating. Additionally, Beckerman's concept of vertical analysis can be directly compared with Burrows' suggestion that processual analysis 'would work with a dense series of instantaneous takes of stages in the musical process' (1979: 248) with the objective of accounting for the manner in which an audience's engagement with music is controlled by the temporality of the live performance.[4]

This description of a 'series of instantaneous takes' bears a striking resemblance to what is also manifested in much dance criticism. As examined earlier, there exists a strong descriptive bias behind the theory and practice of much dance reviewing. One explanation suggests that much contemporary dance, often lacking clear narrative, overriding unity, or established semiology, seems to resist interpretation and evaluation (Banes, 1994: 24–30). As a result, the objective of the dance reviewer is often description and the detailing of what was there and telling the reader what happened. However, inevitably unable to account for each and every instance of the performance reviewers describe what are literally only 'moments' of the production, selected images that are implicitly illustrative or salient of the overall performance.

The danger is that rather than opening up the performance as a temporal event, such presentation of fleeting moments might 'still' the performance: the moments are isolated like photographic stills, the production envisioned as if constructed, cinematographically, from a series of frames. To do otherwise is extremely challenging, but is something that I think is achieved in the work of American dance critic Marcia Siegel. The following example is from a review of 1977 performance by the Trisha Brown Company in New York:

> In 'Spanish Dance' there's a line of women across the space, all facing the same way. The one on the end lifts her arms overhead and begins to walk with tiny, emphatic steps that displace her hips and pump her knees. She gradually activates the next women in line as she comes up behind her, and the next, until all five women are strutting front-to-back, like some chorus line caught in a phone booth. (Siegel 1991: 19)

Here Siegel uses spare but evocative language to create a sense of progress through the narration of movement in time (and also communicating a sense of the space through the body of the moving performer, more on which later). She does this by first re-presenting the moment for the reader, constructing a mental picture that is immediately narrowed to the action of a single performer. Then, as each dancer in turn joins the movement, the reader's mental picture is progressively widened. In a sense Siegel has recreated her own changing focus of attention during the live performance in the mind of the reader, progressively narrowing and progressively widening, as this 'moment' played out in time.

A processual or temporal writing of performance would, therefore, seek to be reflective of the manner in which an audience responds to the temporal flow of a performance. Moreover, there is a sense that what is represented to the reader are the salient moments of the production, central to the nature of the lived experience. The concepts contained in terminology such as 'climax', 'epiphany', 'leitmotif' and 'coup de théâtre' all suggest that it is the identification and perception of distinct and vital moments, as much as the whole, that determines the experience. Such terms reflect the fact that not all moments are equal and that there are certain instances of a performance that define the event, and the audience's experience of the event, as a whole. A coup de théâtre, for example, is defined by Pavis as a 'totally unexpected action that suddenly changes the situation, development or outcome of the action' (1998: 83). In other words, it is a moment around which all others take their meaning and their signification. Our memories, therefore, are grouped around these particular moments, rather than systematic or structured recollection of the event as a whole, and it is this kind of narration of the momentary experience that we might expect to be found in the language of performance reviews.

Unlike the previous sections on video and still photography, this chapter is conducted in the same discourses as its subject (words about words), which makes it possible to illustrate these possibilities further through the presentation of writing of my own. This can be done through adapting, editing and adding to extracts from several of the *India Song* reviews, familiar from Chapter 9, to produce something more deliberately directed to a writing of performative time. As previously discussed, in *India Song* the onstage performers are almost entirely mute. There is one exception to this, one moment when an actor speaks aloud on stage. The following 'review' seeks to evoke for the reader the emotion of this 'moment', conveying the sense of interruption and unexpected feeling provoked by the sudden speaking live on stage:

> In *India Song*, no word of the script is spoken in view of the audience Instead, the action is narrated by off stage voices, which drone from speakers mounted on the arms of an enormous metal fan revolving over and dominating the entire

stage. The unseen voices are recorded on tape, and so too are the stage directions edited into a fast-flowing barrage of detail, which is sometimes echoed, sometimes contradicted by the movement on stage. The characters mime to the half-remembered dialogue, seemingly helpless to escape from their destinies.

A section of the audience is also on the stage: bathed in pervasive sepia light, lounging in soft black cushions, engulfed by the strong smell of citrus and spices in the air. The play shows us the world of European colonials, who languish indolently in the heat of India, only street sounds and music hinting at the teaming life happening somewhere else. Boredom dominates, and lazy sensuality marks the developing sexual obsession of the vice-consul from Lahore for the French Ambassador's wife. A sense of lassitude, of torpor and monsoon dampness descends as the six silent actors drift across the open stage.

Then, in a moment of agonising and unrefined emotion, the vice-consul roars out his huge, frustrated longing in his own voice. This raw live voice cuts through the studied ennui of the characters' world, with visceral impact on the transfixed audience. It slowly drifts into silence; the difference in tone, in meaning and emotion between the cry of real feeling and the lethargy of the unseen voices echoing in our minds as the spoken stage directions sharply interrupt: 'Black.'

This re-visioning of the *India Song* reviews hopes to demonstrate how writing might convey a sense of the temporal experience of live performance. Here evaluation and interpretation are present, but take a backseat to description directed towards inviting the reader to imagine what the experience was like as it happened. And already, additionally, this example shows how the positioning of the writer (and by proxy the reader) in time necessitates awareness of the experiencing body and the beginnings of the writing of performative space.

Writing space

The concept of a unique temporality to live performance is indivisible from the concept of an accompanying spatial dimension: 'now' is always accompanied by 'here'. In performance discourses this understanding of spatiality is associated with several overlapping ideas of presence. 'Physical presence' is a long-standing if frequently under-defined concept used by performers, artists and audiences to describe the tangible (if ineffable) closeness of the performer in time and space during a live performance. Presence is also to used describe the strong sense of the audience's attendance and attention that is felt by the performer in return. As Ian Watson writes,

> stage presence [is] the indirect interaction between actor and spectator, and the connection between the psycho-emotional experience of the actor and its impact on the spectator. (1995: 145)

Other writers also place particular stress on the human performer, evoking a unique communication between the live audience and live performer. Francis Sparshott's description of this relationship is typical: 'in the concert hall even the most introspective performer is playing for listeners who are listening to him' (1987: 89). And Beckerman also stresses the importance of the existence of the performer on stage: 'Eliminate the actuality of man and eliminate theatre, the experience of seeing human beings battling time and space cannot be the same as seeing visual images upon a screen' (1979: 7).[5] The spatial uniqueness of performance, therefore, is rooted in the co-presence between audience members and between audience and stage. Manifestations of this spatial experience can be found in some performance reviews, with a good example again coming from the writing of Marcia Siegel.

Indeed, Siegel remarks in a couple of reviews on the physical relationship that exists between the dancer and the audience. She observes, for example, that the physical awkwardness of a movement makes her physically uncomfortable as she imagines the movement in her body. In one detailed passage she notes how the rhythmically repeated movements of one performance immediately set up sympathetic response in her own body. She then notices what she describes as 'a tiny hitch in the right-left symmetricality of my inner echo' (1991: 80) and realises that her body has responded to a change in the performance rhythm before she was consciously aware of it. The challenge is to render such observations of kinetic transference linguistically, a difficult task that Siegel does occasionally achieve, here in a 1979 review of American Theatre Lab:

> Seated sideways on one hip, [Toby Armour] sharply turns her head and extends one hand in the same direction – and suddenly the room gets about fifteen feet wider. (1991: 72)

What is particularly striking about this example is the concise way Siegel uses physicality to communicate movement and space. It is a sharply observed illustration of the empathetic relationship between performer and audience; the crispness of Siegel's writing allows the reader to share in that relationship. This description draws the reader into the event, evoking space and physical presence through close attention to the performer's body and a sense of process through detailed narration of movement in time – elsewhere in the same review the phrase 'gradually breathed herself into uprightness, and into space' does something similar. The reason that this kind of description of a physical and embodied space of live performance works, and its potential for development, can be usefully elucidated in terms of phenomenological concepts of the body.

Phenomenology and the body

In his writings on phenomenology, Maurice Merleau-Ponty argues that the body is humanity's basic mode of being in the world and that human consciousness is 'embodied' consciousness. Indeed, Merleau-Ponty emphasises this embodied nature of spatial relationships: 'To be a body, is to be tied to a certain world. ... Our body is not primarily in space: it is of it' (1962: 148). In other words, we experience the world in a particular way, in our particularly human way, because we experience it through our bodies. Stanton Garner usefully elucidates these ideas:

> The body is that by which I come to know the world, the perceptual ground against which the world has existence for me; at the same time, it is an object in this world. (1994: 50)

Space, therefore, is human space, scaled against and inhabited by the human body. The body is both the vehicle by which we experience the world and an active object within that experience. Further, the embodied human experience is what Edmund Husserl describes as an 'intersubjective' (or thereness-for-me of others) experience of a world inhabited by other bodies (1960: 92). As David Stewart writes in *Exploring Phenomenology*, 'To be bodily is to exist in a world inhabited by other persons ... one discovers his own authentic humanity only by recognising the humanity of others' (Stewart and Mickunas, 1974: 63). This idea that we experience the world through our bodies is often assumed and often neglected. Indeed, much philosophy presents itself as conducted in an entirely self-sufficient and self-experiencing mind, continuing the western separation of the mind and body.

Now it is possible to take these ideas of intersubjectivity, and the idea of knowledge as embodied in our physical presence in the world, and apply them to the audience's experience of the live performer. In the context of the communal experience of live performance – individual audience members watching a performer in the presence of a number of other people – it is possible to see how knowledge of the event is constructed not only by our own subjective experience, but also through awareness of the subjective experience of others. Peter Brook, for example, suggests this when he describes the audience as 'witnesses' to the event, with the experience being a form of 'commune' (in Melzer, 1995a: 148); while Herbert Blau argues that an audience 'is not so much a mere congregation of people as a body of thought and desire' (1990: 25). Or as Jean-Paul Sartre observes:

> An audience is primarily an assembly. That is to say, each member of an audience ask himself what he thinks of a play and at the same time what his neighbor is thinking. (1976: 67)

Beyond this, audiences not only ask what their neighbours are thinking, but also feel bodily how they are feeling: as Beckerman writes 'We might very well say that an audience does not see with its eyes but with its lungs, does not hear with its ears but with its skin' (1979: 150). Matching this intersubjective experience within an audience is a similar relationship between individual audience members (as part of the collective audience) and the live performer. Audience members, all of whom have bodies and experience the world through their bodies, are able to empathise with the bodily presence of the performer. Indeed, this empathy with performers and awareness of their embodied presence leads to the communication, or feeling of communication, of the senses of others: whether that is thought, touch, taste or smell. In short, it is possible to feel the thoughts, actions, pleasures and pains of other people through an intersubjective empathy with their body: we literally know how they feel and even feel how they feel. Physical presence, therefore, operates through an embodied empathy with the bodies of others. Live performance reaches especially high levels of intersubjective awareness through the directed gaze and collective concentration of the audience and because of the heightened tension of the performance space.[6]

The implications of these ideas for live performance are significant as they bind the sometimes rootless themes of presence, community and charisma to a coherent world view and philosophy. Embodied phenomenology emphasises spatial relationships, particularly inter-human spatial relationships, of the kind that are at the centre of performative presence. The question now, therefore, is whether the bodily and intersubjective nature of the experience informs the language we use to write about the experience of live performance.

Certainly many dance and theatre reviewers have focused upon the physicality of performers in a way that invites the embodied engagement of the reader. Indeed, in *One Night Stands* Michael Billington observes that Kenneth Tynan's criticism concentrates on the acting and the physical attributes of great actors (1994: 8).[7] The same could be said of Billington himself, who also focuses on the physicality of the performers he reviews. In both Billington's and Tynan's reviews it is possible to select passages, sentences and fragments illustrating this often tight focus on actors and their corporeality. This idea exists in the very grounding of their theatrical theory, where acting is the communication of presence to the extent that Billington writes: 'Like all remarkable performers Eduardo De Filippo ... leaves behind an ineradicable physical imprint' (14). He similarly observes that Albert Finney has 'the born actor's capacity to leave behind an indelible physical image' (44). Theatre, for these critics, is the physical and emotional presence of actors; great theatre is to be in the presence of great actors.

Tynan's evocations of actors and acting are splendidly lyrical outbursts: they read as if possessed by the force that propelled the original performance. In Tynan's descriptions, meaning, emotion and action are entwined together in the bodily being of the performer. Moreover, his writing re-presents this presence through the prism of memory recalled: the reader is given access to Tynan's eyes and invited to see as he saw. Illustrative of this is a description of Ralph Richardson as Falstaff in an Old Vic production of *Henry IV* in 1946:

> [He] never rollicked or slobbered or staggered: it was not a sweaty fat man, but a dry and dignified one. As the great belly moved, step following step with great finesse lest it overtopple, the arms flapped fussily at the sides as if to paddle the body's bulk along. (1984: 31)

This passage operates on a number of levels, with the humour both subtle and clichéd as Tynan renders the fat man of comedy with dignity, but at once disallows that dignity. The description calls on the reader's knowledge of the play, yet denies that knowledge and suggests something new; familiarly evoking Falstaff's great belly, but also allowing it 'great finesse'. The language embodies the physical impression even in relation of what Falstaff did not do: the words 'rollicked', 'slobbered' and 'staggered' are dynamic, kinaesthetic of a large and particularly weighty dynamism. This description allows readers to place themselves intersubjectively in the role of the great fat man and at the same time in the role of the onlooker. In this sense of being there, the passage inspires a false or rather imagined feeling of recollection in the mind of the reader. Using embodied language to re-present embodied performers, Tynan's writing evokes a powerful sense of the spatial and bodily presence of the performer.

Billington's own descriptions of actors' performances rarely have the same dramatic intensity of Tynan's writing, although they also focus on physical presence and appearance in a very evocative manner: such as his description of actors as 'shy, gawky, repressed' or 'short-legged, broad-bottomed, crab-gaited and moon-faced' (1994: 86). Elsewhere, detailing Nicol Williamson's performance as Uncle Vanya in a 1974 RSC production, Billington writes:

> he goes brick-red with impotent fury, he makes short, nervous, stabbing gestures at the Professor and essays aimless kicks like a thwarted infant. ... his body straightens, his eyes bulge and you feel confronted by temporary insanity. (59)

Here, as in Tynan's writing, physical appearance is the grounding of analysis. Billington's language seeks to embody its meaning in empathetic detail; in the same review he describes Williamson's 'shambling figure' and 'gangling body' which 'constantly seems a dreadful encumbrance to him so that when he tries to curl up unsuccessfully on the bench you feel that his inordinate frame might snap in two' (59). The review presents this description as both a general appearance and a particular moment: a selected and illustrative instance of the whole. Billington's use of the word 'you' here (and elsewhere: 'if you listen' or 'it gives you') seeks to tie the reader into his seat, inviting us to experience the review as audience to the performance. Similarly, his descriptions of actors often links their physical performance to the physical presence of the audience: 'we, the audience, sit a few feet away', 'as we enter' or 'we realise he speaks the blunt truth'.[8]

Minute attention to the physicality of the performer is also frequently found in the best of dance reviewing, where it points towards the prerequisite of how movement emerges from the performer's body. Edwin Denby, in particular, manifests this often very tight focus on dancers' appearance and movement, with attention to the figure of the dancer occurring again and again in his criticism. As, for example, in a description of Anna Sokolow in 1943: 'Her figure with the small head, the solid neck, the small sloping shoulders and elongated limbs was immediately touching' (1986: 181). Or in a description of Tamara Toumanova in 1944, which goes to some length to detail 'her large, handsome, and deadly face, her sword-like toe steps, her firm positions, her vigorous and record-high leg gestures' (254). And again in an article where the striking emergence of Merce Cunningham's choreography was immediately connected to his physical presence: 'At first he was quite extraordinary because of the sloping shoulders and long arms and long legs' (406). These intense almost intrusive bodily descriptions are clearly a trademark feature of Denby's writing, which is unusual in the intensity and explicitness of his gaze.

With classical dance, interest in the development of balletic tradition leads the critic to focus on dancers' physical interpretation of technical movements. The critic of new dance works also often focuses on dancers' physical bodies, although now for different reasons. Marcia Siegel suggests that modern dance is often closely identified with its creator and their bodily shape and movements. Consequently, in her reviews, Siegel describes the body of the dancer with an attention to detail almost as intense as that which Denby focuses on classical dancers. The difference is that, while Denby might have been concerned with how the dancer affected the dance, the key factor for Siegel is how the dancer determines the movement. Given the range of twentieth-century dance choreography, different parts of the body may now be of interest and different movements performed, calling on a very different kind of physical description on the part of the reviewer. For example, Siegel describes dancer Judy Padow as

> small, with a thin, flat body that shows little of the refinement that usually comes from dance training. She doesn't have that high muscle tonus, that readiness and sense of reserved power you see in other dancers. Her body looks put together with soft, weightless things like marshmallows. (1991: 40)

Described in contrast to the typical dancer, Padow is soft and weightless where Denby's dancers were powerful and weighty; and it is difficult to imagine Denby positively comparing a dancer's body to marshmallows. Elsewhere Siegel describes dancers as angular, even awkward, jarring and harsh. As twentieth-century choreography has redefined dance aesthetics, so in turn is a new dance vocabulary required. However, while the vocabulary may change, the types of words are similar, all calling on embodied knowledge – perhaps we can even feel in our own bodies the idea of being 'like marshmallows'.

Modern dance's deconstruction of tradition, and the movements of traditional technique, means that the writer does not have an expert, technical vocabulary to utilise. Without a technical lexicon, and without set moves to notice, register and assess, the critic is dependent on rendering observation into language: telling what happened on stage. Such descriptions seem to aim for exacting observation, description of what happened without judgement – there is often no set standard against which to judge – and often without interpretation. Yet exactitude is unachievable, the observer's eye cannot be all-encompassing, and the critic's writing cannot be precise. Instead the writing needs to be selective and yet also evocative, a possibility that can be explored through the concepts of synaesthetic language.

Synaesthesia

In *Music as Heard*, Thomas Clifton relates Merleau-Ponty's description of the embodied experience of the world to the specific experience of music. In particular, he discusses how our experience of music is not limited to the auditory but, like the world, is experienced through the whole body: that is to say synaesthetically. Synaesthesia refers to how emotions or stimuli to one sense can prompt responses in another. In other words, perceptions resulting from sight, sound, smell or touch are not isolated experiences but phenomena unified by the human body. Clifton notes that, 'Synaesthetic perception forms an important part of Merleau-Ponty's phenomenology of the body, as a "general instrument of comprehension"' (1983: 65).

Synaesthesia, therefore, describes the association of different sense experiences. Although as a medical condition in extreme cases it is potentially disabling, Peter Morris and Peter Hampson suggest in *Imagery and Consciousness* that 'Mild synaesthesias are reasonably common in most individuals, certain colours are often described as warm or cold, and sounds as bright' (1983: 110). Indeed, weak colour associations are among the most familiar of synaesthetic constructions: red is hot and stressful, green is cool and calming. Many common phrases also illustrate how bodily responses to one stimulus are linguistically or experientially linked to other sense perceptions. For example, when we say something is 'cringe-making', 'eye-watering', 'heart-rending' or 'blood-chilling' we indicate how we instinctively describe our bodily experience of sensual phenomena. It is open as to whether this is a form of weak, experiential synaesthesia, or of purely linguistic construction. However, either metaphorically or physically such phrases do detail our whole bodily reaction to what might be sensations empirically limited to just one or two external perceptions.

This concept is especially relevant to the discussion of performance, as the arts seem to encourage synaesthetic comparisons, because they inspire their audience to heightened levels of cognition. Certainly, commentators often employ what is effectively synaesthesia to describe the aspects of performative experience that defy explanation. Barba talks about the way an actor's body feels the tension of the audiences (in Watson, 1995: 144), Blau suggests there is a physical force to the audiences' gaze (1990: 6), while Beckerman describes how an audience feels through the skin (1979: 150). While the synaesthetic bodily experience is something often remarked upon in relation to a range of arts it is particularly interesting in relation to performative liveness.[9]

Ideas of co-presence describe how we experience performance through the spatial simultaneity of the human body. As such, we attend performances in and with our whole bodily person: we are all there, not just watching and listening. Synaesthesia represents linguistically how the whole body and all the senses are engaged in the experience of live performance. Potentially, therefore, the employment of synaesthetic language to discuss performance begins to create an 'embodied' writing that places the reader in the bodily, physical and spatial location of the performance. An 'embodied' language would be one where the choice of words and the attentions of the writer are directed by the perception of experiences as physical and human; embodied language communicates this bodily experience of the world. Indeed, much language is always going to be embodied, as its origins (as demonstrated by the etymology of many words and phrases) are already rooted in human and hence embodied experience. In the specific context of the representation of the experience of live performance, these possibilities are worth exploring further, both theoretically and practically.

Clifton suggests that the spatially or bodily oriented terms used to discuss music – terms such as high, low, rounded, pointed, bright, dark, bouncy, rough, hollow – are not merely metaphorical or allegorical, but are the pointers to the synaesthetics of perception. Taking this further, the use of such spatial terms in music criticism is more than just a clue to synaesthetic perception, but is also the result of the experiential perception. It is language chosen to discuss an experience because of the bodily nature of that experience. Clifton persuasively argues that we describe music in terms of textures, space dimensions and bodily-located words because that 'is what we experience when we hear durations, registers, intensities, and tone qualities.' (1983: 69). For example:

> the sound produced by an oboe in its middle register is usually described as somewhat thin, nasal, rough, and slightly hollow. But this is not altogether accurate. Rather, these words are descriptive of our bodily behaviour: we have adopted an attitude of hollowness, thinness etc. (1983: 68)

Clifton also presents other examples: we say sky blue is restful because our body has adopted a mode of restfulness, we say a movie is edgy because our body has adopted a mode of edginess, we say music is bouncy... and so on. These, argues Clifton, are not stimulus impinging on the body, but effects produced by the body, without which the responses would not exist. It is not the sound of the oboe in and of itself that makes it thin, but rather the bodily perception of the sound. The restfulness of sky blue is not in the colour blue but in the bodily experience of that colour. With some reservations, it is useful to see such words as more than metaphorical, just as the experience of presence in live performance is also more than merely metaphorical. However, whether metaphorical or allegorical, such terms do have particular evocative use in their relation of experience to the human body. Consequently, awareness of synaesthetic comparisons in language could provide us with a language that accounts for the spatial and bodily experience performance.

A good illustration of a music critic who uses such terms is Andrew Porter, who frequently details music in terms of 'texture' – contrasting adjectives like 'thick' and 'sticky' to verbs like 'rippled' and 'purled'. This kind of imagery provides music, synaesthetically, with a physical, embodied form and appearance as physical as that of dance or theatre.

Such synaesthetic description of music recurs in many different forms in Porter's writing, such as a review of Vladimir Horowitz that first bends music to shapes – 'sharply etched, incisively sounded, quirkily phrased' – before moving on to human actions: 'boisterous', 'jerking' and 'jittery' (1979: 41). When such human actions are used to describe music, is the implication that the music itself is boisterous, or that it engenders a sense of boisterousness in the audience, or a combination of both? The possibilities are added to by looking at the similarity of the language used to described music, and the words Porter employs to detail the listener's response. Porter often describes audience responses in terms of violent physical reactions, on one occasion 'good enough to leave the listeners rapt, awed, exhilarated, trembling' (1988: 53). When reviewing music by Mahler, which seems to particularly inspire such descriptions, the meaning of the phrase 'shattering' blurs between the music and the listener. Porter describes the music of another performance as 'despairing, insecure, even hysterical' (1988: 60) – or is that a description of the listener's response? Such descriptions allow readers to imagine themselves into the state of the audience, as experienced by the reviewer. Additionally, they also, because of Porter's extension of such terms from audience response to the music itself, invite the reader to imagine the music as well. Indeed, his writing is an interesting reversal of a statement Porter makes in one of his reviews that 'forms and feelings and thoughts have all become sounds' (1979: 426). In contrast, in his writing sounds become thoughts, feelings and particularly forms.[10]

The employment of synaesthetic comparisons, therefore, effectively reflects (or rather embodies) our bodily experience of music, of performance and indeed of the world. Taking this further, Arthur Koestler and Stanley Burnshaw, both with interests in linguistic theory, provide crucial suggestions about what effect such language then has on the reader. In *Insight and Outlook*, Koestler describes the value of synaesthetic comparisons in terms of the opportunity they provide 'for sympathetic projections of emotions, for identification with other selves' (1949: 319). Koestler continues:

> It is obvious that such 'synaesthetic' metaphors greatly facilitates the sharing by the reader of the teller's vision, as more of his sensory fields are mobilized to participate in the experience, which thus becomes multidimensional, fuller and richer. (1949: 320)

It would be possible here to replace Koestler's 'sympathetic' with the phenomenological concept of 'intersubjective'. The projection of synaesthetic experiences between reader and writer is possible because of the shared bodily experience of the world: the relationship is intersubjective. Burnshaw presents somewhat similar arguments in *The Seamless Web*, where he clearly describes (without using the word himself) synaesthetic reactions to art and the world: 'the entire human organism always participates in any reaction' (1970: 10). For Burnshaw this also includes reading, for in discussing responses to poetry he suggests that the reader 'cannot help but read into the words images of his own body' (1970: 268). Similarly, the use of synaesthetic, intersubjective language in reviews would potentially produce in the reader the cognate linguistic experience of the live performance. To continue with an earlier example by way of demonstration, the description of the oboe as 'hollow' not only represents the writer's bodily experience of the sound but also invites readers to enter into that experience: indeed, it encourages readers to have that experience in their own bodies.

The significance of these ideas to the language of reviews should be evident, suggesting a method that unites language not just with experience but also with the nature of that experience. We can write about our bodily experiences in a language that allows readers to think themselves imaginatively into the experience: an experience that the body grounds in the physical space of the live performance. Additionally, such language use (following Clifton) is not an attempt to escape intellectual rigour or to avoid dealing with the thing itself. Instead, it is non-allegorical and reflects the bodily nature of the original experience.

At least, that is the possibility. And with this discussion taking the form of words about words, as with the earlier example with the writing of time, it is worth illustrating these possibilities further through my own writing. The following re-visioning of the *India Song* reviews seeks to do this by evoking for the reader the physical space, embodied presence and synaesthetic experience of the performance:

With their action narrated, sometimes inconsistently, by unseen voices the performers in *India Song* do not always seem entirely present in the space they occupy. They drift, silently, across the open stage, often appearing to move with unseeing eyes while mouthing to dialogue that is not being spoken. Surrounding the performers, part of the audience is seated on-stage, drenched by the hot lights and sudden inky darkness, lolling back in deep-black cushions, assuming the same attitude of torpor and lassitude than infests the characters. The stage trusts out into the auditorium, sickly-yellow sodium streetlights hanging over the entire audience, placing them in the same colonial corral of the play. Scent drifts round pervasively, and sounds filter in from an India outside this isolated and enclosed world.

At one stage, the men swarm lazily to an expatriate party, desiring to dance and flirt with weary sensuality with the French Ambassador's wife. They appear in evening-wear, the vice-consul of Lahore glowing bright-red in his pristine white suit, his bald-head dripping with perspiration that enforces a deadening lassitude on all his movements and emotions. The buttoned-up restraint and heavy formality of these European's clothes collides violently with the temperature and temperament of the world in which they find themselves.

The contrast between real sweat, the physical reaction to our actual environment, and hollow society-restricted emotions emerges when the Ambassador's wife dances with her husband, flowing purposelessly around the stage. The Ambassador is played by a dummy, also dressed in evening-wear, also sweating profusely, also dumb, with that same blind stare shared by all the characters. Throughout, the evident effort of the studied appearances compromises any degree of gracefulness or sophistication, brutally undercutting the public façades. It is only when the vice-consul's longing breaches the constraints of politeness and society that any of the characters seem to come together as people, eyes finally provided with sight, limbs with muscle and purpose, and mouths provided with speech.

This example seeks to demonstrate how a sense of physical, embodied location and space might be constructed through language. As Della Pollock argues, such writing seeks to express experience in which 'The writer and the world's bodies intertwine in evocative writing, in intimate coperformance of language and experience' (1998: 81). A spatial and embodied writing of performance would write about those elements of presence that are vital to the experience of live performance, thereby representing the performance as live and inviting the reader to experience the event by proxy of the writer's own embodied perceptions.

A poetics of liveness

The chapters in this part have examined the conventions, circumstances of production, structures and grammars of newspaper reviews in close detail, highlighting examples where I believe language begins to inscribe the performance in writing. Discussion has explored trace, established and potential elements of language that instigate a reverberation of the live experience in the mind of reader. The proposals have been speculative, drawing upon ideas of phenomenological perception and identifying the power of description when charged with meaning through bodily experience.

A writing of live performance can be summarised as one that focuses on and communicates the experience of performance as an event occurring in a uniqueness of time and space, with suggestions partly made through identifying characteristics of successful performative writing already present and utilised in criticism. Such evocative description is particularly in evidence in the work of writers such as Billington, Denby, Porter, Tynan and Siegel, who at their most evocative present readers with the opportunity to think themselves imaginatively into the location and moment of the performance. It is also possible to identify an echo of the writer's experience of performance in contemporary reviewing more broadly, as represented here by the *India Song* press coverage. However, such evocative description and representation that does exist is often relegated behind other (and indeed very legitimate) competing demands and preoccupations of performance reviewing – especially for evaluation, which can be seen as central purpose of reviewing as a form.

Influenced by the fact that this analysis is in the same semiotic system as its subject (words about words), some of this discussion has taken an unusual, self-illustrative and even quasi-prescriptive tone: calling for better, more evocative, more responsible and more self-aware writing about performance. In drawing this analysis to a close it is worth revisiting the four elements of descriptive writing that have been identified as particularly successful in representing performance – the evocation of intersubjectivity, the suggestion of multisensory synaesthetic responses, the embodiment of space and momentary analysis – which together form a kind of poetics for the writing of the live experience.

Evocation of intersubjectivity

The concept of intersubjectivity, at its basis, draws attention to the level on which our experience of the world is a human, bodily and intrinsically sharable experience. Intersubjectivity roots ideas of empathy, representing an invitation and ability to see what others see and feel what others feel. This utilisation of this concept here is several-fold, borrowing in part from the long emphasis on physical presence in performance theory and practice. Additionally, and vitally in this context, there is potential for an intersubjective relationship between the reader and the writer, and hence by proxy between absent readers, present audience (as represented by the writer) and the experience of performance.

To facilitate the development of a potential reader/writer intersubjective relationship, readers must be invited to imagine the performance via the body of the writer – to metaphorically take their seat at the event. This requires the creation, in language, of a sense of an experiencing subject: the location of the experience in a specific mind and body. This is achieved by reference to an individual or collective gaze and presence, constructing a representation of the performance as a human scaled and experienced event. This resists possibilities of abstraction and detachment, which can result from writing seemingly originating from an anonymous and unbodied 'expert' voice. An intersubjective writing denotes the experience as one of people and with people and therefore a definitively live experience.

Synaesthetic experience

The establishment of an experiencing subject implies an embodied, human and hence sharable perception. Language can encourage such communication through the writing of the bodily experience as a synaesthetic experience. Audiences witness performances through their eyes and ears, yet they experience it with their whole bodies. Synaesthesia describes the transference of sense terms from one sphere to another, a state where sensations produced by one stimulus are experienced (or described as if experienced) by other sensations. Moreover, by using embodied synaesthetic descriptions writers present their own bodies as the medium through which readers can access the experience for themselves. By translating emotional and intellectual responses into embodied reactions, readers are empowered to intersubjectively access that experience with their own bodies. Combined with the intersubjective invitation, the inclination is once more to consider what it would be like to be there.

More generally, synaesthetic comparisons provide a broader and more multidimensional sense 'picture' of the event being described, establishing a rich and full impression of the live performance. Synaesthesia offers the opportunity for the extension of sensual perception and the 'thickening' of the reader's descriptive experience. These values – multidimensional, richer, fuller – are ones associated with, defined by and defining liveness. Synaesthetic writing seeks to affirm perception of the presence of human performers and audiences at the performance, marking it out as an event experienced live – and therefore involving all the senses.

Embodiment of space

Awareness of a unique spatial experience of theatre and dance is a recurring theme in discourses of performance. Language can evoke a sense of this performative space through the close descriptive detailing of the physicality of the performers. The performer occupies space by breathing, by moving, by looking, by standing; creating a very familiar sense of human-scaled and human-occupied space. Engaged readers are then able to imagine the performance through their own bodily awareness, thinking themselves into the movement, actions and emotions, into the body and into the space itself. Such embodied language resists abstraction, which can result from writing that undermines such qualities through presentation of purely intellectual perception. It also enforces upon readers the physicality of the performer and thus links the description directly to the fundamental experience of presence in live performance.

Momentary analysis

The most dynamic descriptions and evocations of space are those that envision it as existing and changing in time. Audiences experience a performance as a continuous flow, but it is often perceived as if it were a series of interlinked moments, some of which stand out from the stream as particularly important. These are the moments around which our understanding and experience of the event is structured. This is the case both in memory – recollecting an event now finished – and in the synchronic experience. Peaks, turning points, repeated motifs, climaxes and tableaux are all perceived, as they occur, to be of particular importance to the experience as a whole. In relation to photography, Chapter 7 explored how moments captured by the camera are privileged, perfect or decisive. Just as still photography has sought out these pivotal moments, so can written representations. The challenge with still photography is to make the static image refer to moments occurring before and after, to encourage in the viewer the imagination of change and process. In its very different form, the challenge is the same for the linguistic representation of live performance.

As a fixed entity writing can present a static, omniscient impression of a performance, transforming it into something all visible at any one moment. Momentary analysis resists such temptation, instead seeking to re-create in language the structural importance and experience of the individual moments: not unitary instances but changing processes. Momentary writing demands closely detailed description of the performance as occurring and changing in the temporal experience. The impression of process that is subsequently inspired in readers roots the description in time and therefore encourages imagination of the performance as occurring over time.

Writing cannot be mimetic to the direct appearance or fact of performance. It cannot be objective or complete, but instead it must always be shaped by the writer's own opinions and subjective responses. However, through the utilisation of the tools and engaged level of attention outlined in these four techniques, writing can be evocative, can represent the writer's sense of being there and as a result provide access to their experiential knowledge of performance. It is in this re-presentation of the observer's experience that the usefulness and purpose of writing about performance rests. Such writing neither simply asserts the writer's opinion, nor passively copies down surface appearances, but instead provides us with an engaged 'audiencing' of the experience. That is, it provides us with an opportunity of seeing and of knowing the performance as an event for an audience – as an event that is not complete in itself until seen, known, engaged with and interpreted by an audience. It is in terms of such audiencing, and the statement therein that performance is experience, that the profound transformational significance of representations of performance is located.

11
The Representation of Live Performance

The concept of disappearance articulates the live arts as actively resistant or impervious to representation, with this transience prominently constructed as a marker of the unique character, force and meaning of performance. At the same time practitioners and theorists alike also demonstrate a powerful preoccupation with questions of documentation, prompted by a fear of forgetting and the need for performance to exist within wider cultural and academic discourses and conversations. Partly contradictory, partly mirroring, each of these discourses of documentation and disappearance drives and motivates the other. They are mutually entangled, mutually dependent, with ideas of disappearance depending on the continued existence of some manner of trace or presence and acts of documentation always involving some degree of fragmentation and absence. Together these two themes underwrite much conceptual discussion and thinking within performance studies.

The play between these two positions deconstructs any possibility of stability. The formulation of the live as 'disappearing' can be challenged, not least by the very existence of fragmentary traces that mean that performance never *only* exists in the now of its live creation. But while performance certainly does exist within systems of representation, reproduction and exchange, it does not do so easily or unproblematically. Unlike non-live, technologically originating performances – which are not so much re-performed as re-played – each repetition or representation of a live performance marks it as different. Produced in the context of this central dilemma or problem, representations of performance possess a particularly contentious position in relation to their subject, always loaded with ideological and ontological significance as well as artistic and even moral implications. Continually examined via the media, interests and abilities of its representations, performance is continually presented – even continually present – in forms essentially other than itself. Consequently, the methods and media of representation have a powerful ability to shape the cultural constitution and meaning of performance.

The ontology of performance oscillates between presence and absence, between remembering and forgetting, between documentation and disappearance. With this in mind it is necessary to re-think the nature of the point located within and constructed between these positions. If performance neither disappears, nor fully resides in its documentations, then it seems appropriate to think of the continued cultural manifestation of performance as located somewhere in the space and time between. This is a space of extra-performance existence and non-existence: consisting of traces, fragments, memories, forgettings, half-truths and half-lives; consisting of representations that contain something of the thing itself, but which are not the thing itself.

Certainly, unable to be continually *there* in the moment of the performative now, the space between documentation and disappearance is in many ways the primary and only enduring site of our knowledge and understanding of performance. As a result these representations are a site of ideological and sociological significance, where the experience, perception and valuation of the thing itself can be seen and known. With this in mind, this book has explored how and what it is possible to know, see and say of performance through its representations. This last chapter draws this discussion to a close by reflecting on the possibilities of a methodology of analysis that employs these representational traces as an active, interrogative fulcrum through which to know performance.

The act of representation

In *Languages of the Stage*, Patrice Pavis suggests that the basic question to consider in relation to performance documentation is not how to conduct it 'but for what purpose' (1982: 112). To a degree this is certainly correct: as the envisioned function of any documentation – to aid interpretation, to enable revivals, for marketing and publicity and so on – goes a long way in determining the nature of the activity. And such narratives of use have been central to much of the foregoing analysis. However, in concentrating on the rational and deliberate purposes of documentation Pavis neglects the more fundamental and immediate need for representations, which is to enable the very existence of performance within cultural exchanges of ideas, meanings and experiences. Representations of performance are by their nature enduring and reproducible in a manner that the performance itself is not. It is the very temporal and spatial singularity of performance, the fact that it disappears as it becomes, that necessitates its representation. Whatever the specific purpose or commissioning function, all representations of performance result from the same original need and end with the same eventual use: to allow us to see and know something of performance outwith performance itself. The overt purpose and intent of the documentation, therefore, is less significant than the method and manner of its communication.

Writing about the issues raised by this debate, Franc Chamberlain describes the documentation of practice as presenting 'traces left by performance, which are aesthetic objects in their own right, yet remain mute about the performance which produced them' (in Thomson, 2003: 162). Yet while concurring that documents are objects that need to be considered in their own terms, the chapters in this book have explored precisely how such traces do indeed speak for and of performance. Any act of representation is intended to invoke an image, impression or sensation of something that is absent. In doing so it involves active choice, the execution of selection and omission, and the conscious use of technique, all of which can subsequently be employed as a prism through which to examine what was valued, remembered or understood about the performance: to analyse what the performance was constructed to be. As Egil Törnquist writes, 'all notation is interpretative because of choices in what is noteworthy' (1991: 22). Similarly Jonathan Miller's comments on the video recording of theatre are applicable to all methods of performance representation:

> If called upon to copy [a performance] we would copy what we thought was important, and in that very act would adapt the production even if we had agreed to the idea that copying meant producing something indistinguishable. (1986: 52)

It is also useful to adopt the language of still photography here and think about how the camera isolates and freezes a single moment of time and action. What this moment is of then assumes a critical importance: it becomes special, privileged, subject to our gaze and examination simply by fact of being photographed. Yet what this moment is of is not accidental, nor objectively found and recorded, but the result of choice on the part of the photographer. As a result, choices in what to frame and what not to frame become constructive of a particular view of the world and an act of interpretation. As John Berger writes:

> Every time we look at a photograph, we are aware, however slightly, of the photographer selecting that moment from an infinity of other possible sights. ... The photographer's way of seeing is reflected in his choice of subject. (1972: 10)

Similarly, Douglas Harper writes that 'From the phenomenological perspective, photographs express the artistic, emotional, or experiential intent of the photographer' (2000: 727). What the photographer has decided to show, and how they have chosen to do so, communicates particular perceptions, values, meanings, interpretations and 'ways of seeing'. In other words, photographs do not only (and imperfectly) allow us to see something of the world, but also say something about the meaning and value invested in the thing depicted, both in terms of individual responses and experiences and more widespread cultural perceptions and understandings.

Applied to the representation of performance, this description can be extended out from photography to include other media and activity of representation. What a critic decides to include in a 500-word review, what a curator decides to keep or throw away, what an academic thinks is worthy of research and documentation or how a video recording frames and edits a performance are all the result of choices made according to both the media, aesthetics and capabilities of the representation and the values, meanings and ideologies perceived in the performance. They all articulate ways of seeing.

Representations of performance, therefore, do more (or less, depending on your point of view) than merely allow us to see performance. Instead, with all representations, the methods, medium and interests of the presentation also begin to constitute a distinct identity of their subject. They all begin to speak of the performance that produced or inspired them. Indeed, we might similarly think about how which memories are retained and recalled in the audience's recollection also speak for the meaning and nature of the performance remembered. In this, representations are a form of audiencing, a form of watching, seeing, experiencing and making sense of performance.

Moreover, representations can be considered a kind of audience interaction in the manner by which they make that experience observable and knowable through the process of transformation into another medium. For each act of representation is transformational: mediating, changing, mutating its subject as it is transposed from one medium to another. In terms of the discourses of documentation and disappearance this mediation is frequently articulated in terms of betrayal, involving an inevitable loss and reduction of either the essential ontology of the performance or in terms of failing to produce a complete and accurate documentation. The result is a relationship fraught with compromise and suspicion: compromise over what is lost and what is recorded; suspicion over whether the documenting medium is serving its subject or its own values and specificity; doubt over whether the essential nature of the original artefact has been lost somewhere along the way. In response to such dilemmas, Gay McAuley argues that the objective of documenting performance should be to capture 'as fully as possible the pertinent elements of the theatrical experience, without excessive distortions due to the nature of the recording medium' (1994: 187). The proposal contained within this book, in contrast, is that our understanding and knowledge of what was pertinent about the original experience is significantly shaped and directed by the particular specificities of the various recording media. Indeed, it is the very prism of these medium specific transformations (not least amongst which is the very physical endurance that is alien to live performance) that allows us to see in the first place.

All representations of performance can be considered as discourses, governed by the qualities of the medium, by tradition, theory and practice through which the experience and appearance of their subject is made manifest. It is self-evident that each medium is limited by its own inherent specifities and characteristics: what they can say and show of a performance is limited to how they can say and show it. A still photograph cannot record movement, and although it may communicate some idea of movement through visual representation it can only do so photographically, not performatively. A review can only operate through linguistic description and evocation, rather than mimetic reproduction. Similarly, how each medium of representation is employed and interpreted is partly the result of its history and aesthetic of use. One example of this is the strong cultural narrative of photographic revelatory authority, and the challenges presented to this by actual narratives of production or by digital technology. Another is the existence within dance criticism of a strongly anti-interpretative aesthetic, leading to an emphasis on description of the sensual experience rather than analysis or intellectual interpretation. In these manners, each medium operates within its own history, aesthetics and language: established conventions and codes of communication that construct an aesthetic narrative.

Our analysis, therefore, should not attempt to do the impossible and compare the representation with the original, judging according to criteria of accuracy and truthfulness – or rather it should not be only to do this. Instead we should explore the meaning of representations in the space that they construct for themselves. Representations of performance are partly a matter of what can be communicated (given the nature of the representing medium) and partly a matter of choice, objectives and the cultural valuation of the subject. All documents and traces of live performance must be considered as presenting cultural, political and artistic understandings and values of their subject *and at the same time* be analysed with an understanding of the abilities, traditions and objectives of the representing media. As a result it is possible to apply a visual, verbal, media-specific discourse analysis to a consideration of how these representations constitute live performance and how they communicate transience in permanence, movement in stillness, presence in absence, disappearance in documentation.

The analysis of representations

In relation to a range of activities and media this book has set out to know performance through its representations, presenting and utilising different reading and viewing strategies, thinking about questions concerning their content, their circumstances of production, their narratives of use and the aesthetics and specifities of each particular medium. To an extent this is what performance analysis in research and academia frequently (if implicitly) already does involve: as a performance, unavailable live to new audiences, is studied and explored in its absence through its various representations. Yet too often this is conducted without methodology, either hoping through unthinking trust or neglectfulness that the document provides a neutral window onto the thing itself or, more frequently, conducted with narrow distrust of the transformative interpretation presented to us and rejection of any new kinds of knowledge and vision that might be present in the representations themselves. This is to neglect the potential reward of actively looking *at* representations of performance, rather than maintaining the conceit of looking through them.

At this point it is again worth stressing that, except by being *there* in person (and even then only for and at that moment), performance is only available to study, to know and even to think and talk about through its representations. The result is that the enduring life and lasting cultural meaning of performance is found in the space between documentation and disappearance, found and made manifest by the representations of performance. As a result the enduring representations of performance inevitably have a certain power and authority over their subject, directing and shaping how we see, understand and know the thing itself. Indeed, Amelia Jones put this relationship in absolute terms, arguing that the meaning of live performance is retrospectively formed by the documentations and interpretations of historians and critics (1997). Similarly, Eugenio Barba writes of theatre practice that 'what really matters is what will be said afterwards when we who worked at the task are gone', with this afterlife of the performance experience determining its enduring and eventual significance (1992: 77). In terms of the power relationship between performance and its documentations, we might also see something of Rosetta Brookes' observation that when an 'event in its reproduced form becomes socially more important than its original form, then the original has to direct itself to its reproduction' (in Elwers, 2005: 160). More certainly, as Derrida writes, 'what is no longer archived in the same way is no longer lived in the same way' (1995: 18), a comment that can be transposed to apply to each forms of representation considered in this book.

This manifest power of the representation to shape and frame our knowledge of performance produces a suspicion and fear of documentation. This is particularly so if the unique attribute of performance is considered to reside in its qualities of liveness, which is not least constructed as disappearance and therefore antithetical to documentation and the 'commodifying clutch of representation' (Heathfield and Quick, 2000: 1). This is a tension articulated time and again in performance studies, including by Helen Freshwater who responds to the problematic nature of the performance archive with the statement that 'No amount of video, documentary recording, or personal testimony can capture the ephemerality of performance. Something will always be lost in translation' (2003: 754).

This rather wistful desire to capture the ephemeral is precisely the meeting places of the simultaneous valuation of both documentation and disappearance, a contradiction that seems unsolvable until we realise that it is the act of translation that is vital. Indeed, it is precisely because theatre and dance are by necessity approached through their various documentations – particularly for the purposes of study or marketing promotion, and particularly on video, by photography and in writing – that it is vital to explore how these representations construct the absent live event. So rather than being regretted, this translation should be seized upon and considered an opportunity for analysis, using the transformational nature of each representation as an act of audiencing that says something of the cultural meanings and perceptions invested in the thing itself.

With its necessary choices and active transformations the effort involved in the act of representation is crucial. The attempt to account for and communicate the experience of something demonstrably other, something intangible, ineffable and ephemeral requires conscious choice, effort and interpretation. In writing about the transience of live art, Adrian Heathfield observes that

> Performance, as an ephemeral event, seems to offer little to tell. It disappears fast and leaves the scarcest of traces for historical record. You really have to be there and even then that's not enough; the event itself is hard to fix in consciousness, in memory and in writing. (2000, 105)

Here Heathfield is precisely right that simply being *there* is not necessarily enough to fully grasp and appreciate the experience. Instead experience often needs to be followed by reflection and by articulation and externalisation, invariably in some form of representation of the experience – typically spoken, often written but also visual or artistic. To represent performance, to document, is also to audience performance. In this manner the act of representation does indeed do much to determine the meaning of the thing itself; in this manner the representations do indeed speak for the work itself. While the primary, sensual, bodily and ineffable experience of being there is not to be undermined or dismissed, without the attempt of representation, of articulation and externalisation, the work itself does not merely disappear but also, to a significant extent, remains mute. Consequently, the analysis of attempts to document, record, remember and thereby represent live performance needs to be placed at the centre of critical considerations of the performing arts, for it is through such representations that we give performance voice.

Notes

Introduction

1. In this context 'discourse' is used to describe the interconnected conversations, vocabularies and writings that group around any particular subject or thing. Indeed, such discourses go a long way to defining and delimiting a subject as a subject. Consequently it is possible to employ the conversations and exchanges that exist about a subject or experience as a forum through which to explore what is understood and valued about it, how it is perceived and what it is culturally constructed to be. This does not necessarily mean that what something 'is' is entirely socially constructed by discursive exchange ('to be is to be perceived'), with the actual phenomenon irrelevant. Instead discourses shape perceptions drawn from the world rather than from nothing. Nonetheless, it is predominantly and significantly discourse that determines the meaning and value of a phenomenon as a *cultural* entity. It is this analysis that this book sets out to conduct, widening the concept of discourse to also consider non-linguistic representations of, on and about performance and considering the values, identities and meanings that they construct and communicate about their subject.
2. This is a description relevant not only to performances of the past but also to performances of today. For even after having seen a performance only yesterday, having been *there* in person, the performance itself is no longer available to us in and of itself but only through its various documentations or through our own doubtful memory and the mediating impacts of trying to externalise that memory through writing, talk, drawing and so forth. The consideration of memory as a distinct (and indeed ideologically loaded) site of performance representation is continued in Chapter 3.
3. See in particular the work, discussions and conferences centred around the Practice as Research in Performance (PARIP) project at the University of Bristol, www.bris.ac.uk/parip
4. Nonetheless, this use of the word 'representation' remains a little awkward. However, there is no satisfactory or agreed upon terminology that accounts for the range of media, notations, reflections and activities that allow us to know performance after, beyond and outside of the thing itself. Although also used in this book, the more familiar word 'documentation' is equally problematic, suggestive of a certain kind of intention, chronology and mimetic directness that is not always appropriate. Other possibilities – such as detritus, trace and afterlife – also suggest themselves, are used at various points here, but are often evocative rather than analytically sufficient. Similarly with Eugenio Barba's suggestion of the Norwegian word 'eftermaele', which he extrapolates as 'that which will be said afterwards' (1992: 77). Amongst all these possibilities the phrase 'representations of performance' seems to most satisfactorily describe the mediating, transformative and interpretative relationship that is the interest of this investigation. The awkwardness of the terminology is itself

an appropriate reminder of the ontological tangle – the contradictory documentation of the ephemeral – from which these extra-performance manifestations emerge.

1. Documentation and Disappearance

1. Although important differences exist in the nature and experience of dance, theatre, music, performance art and so on, this book is primarily interested in the similarities that they share as forms of live performance and within discourses of liveness. These similarities are demonstrated not least by the recurring prominence of ideas of transience. Consequently, the word 'performance' is used broadly to cover the whole range of live arts.
2. Auslander does not so much initiate debate about the relationship between the live and non-live as crystallise ongoing questions and conceptualisations, drawing together in particular ideas from Walter Benjamin (1970) and Jean Baudrillard (1983) and also Steven Connor (1989), Roger Copeland (1990) and Steve Wurtzler (1992). Areas of this discussion are pursued further in Chapters 5 and 6 in relation to video representations of live performance.
3. Indeed, in a slightly different context, Eugenio Barba explicitly states that the spectator 'does not consume' performances (1990: 97), which instead continue and evolve both in future performances and in the audience's memory.
4. Such birthday celebrations are the exception, as we tend not to number performances of a production to differentiate them from each other, as occurs with Super Bowl finals or film sequels, which are manifestly different entities of a much more significant order. Moreover, the labelling of the first night or premiere as distinct from subsequent performances marks the concept of 'firstness' as evidently important even in the context of the 'unrepeatable' live performance. Complicating this issue, previews now often precede first nights, which are consequently no longer literally the first public performance of a production.
5. Actor Sîan Phillips neatly elucidates the complexities of the performative now in her autobiography, where she writes of how she loves being in a production run and steering a performance through the different circumstances that occur each night. As Phillips writes, 'Each night is different and not different' (2001: 197). Elsewhere, Douglas Rosenberg talks about live dance as 'corporeal reproduction' writing that 'the "liveness" of a dance is in the resurrection of the original [...] Each performance brings the dance back to live' (2001).
6. The recent release on CD of British Library Sound Archive recordings of live performances at the Royal Shakespeare Company provides another illustration of this mutual entangling of discourses (and valuations) of documentation and disappearance – this time from within the cultural sector and arts media. For example, actor Antony Sher comments that 'They say that part of the magic of theatre is that it is transitory, it exists in the present, and then it is gone forever. But there is also magic in discovering that we can retrieve some of it.' Similarly, David Oyelowo observes that 'There is joy and sadness in the fact that, as a theatre actor, your toil on that stage will live on only in the people's hearts and minds' (Royal Shakespeare Company, 2005).

7. These issues crystallise when requirements of copyright law confront the problems of dance documentation. Until the second half of the twentieth century, dance's status as non-'literate' performance was embodied in its almost total non-existence in copyright. Instead, the law considered dance under regulations existing for theatre, effectively protecting only the dramatic content of narrative ballets. The law provided no protection for the actual choreography. To copyright something, it had to exist on paper: one had to 'fix' it, as the script was seen as fixing spoken drama or the score as fixing music. This problem was not addressed until the emergence of the Laban and Benesh notation systems, as a result of the recognition of which American law changed in 1978 to give specific recognition to choreography (Guest, 1984: 181–3; Johnson and Snyder, 1999 and Van Camp, 1994).
8. Indeed, Auslander suggests that our contemporary cultural experience dictates that the live is always perceived in relation to the non-live, and that historically the very idea of live performance has been constructed by the concept of the mediatized and is a usage dating back only as far as the 1930s and the development of relatively high quality recording and producing techniques in various media (1999: 52–3).

2. Archives

1. The central accounts in this challenge to archival authority and historical singularity include Roland Barthes' 'The Discourse of History' (1981), Michel de Certeau's *The Writing of History* (1988), Jacques Derrida's *Archive Fever* (1995) and Michel Foucault's *The Archaeology of Knowledge* (1972). A recent and at times polemical champion of the postmodern approach to history has been Keith Jenkins (1999).
2. Also significant is the emphasis on the active and political selectiveness involved in any act of archival collation. Michel de Certeau, for example, describes an archive as a place that is produced by an identifiable group sharing a specifiable practice of organising materials (1988: 76), something often resulting from distinct national remits and narrative (see Brown and Davis-Brown, 1998). Jacques Derrida similarly affirms the centrality of questions of politics and power of archives, writing that 'There is no political power without control of the archive, if not of memory' (1995: 4).
3. As a direct result of this perception, many organisations have been established to directly counter the problem of the disappearance of dance, such as the National Initiative to Preserve America's Dance, which operated from 1993 to 2001 with goal to 'foster America's dance legacy by supporting dance documentation and preservation as an integral and ongoing part of the creation, transmission and performance of dance' (NIPAD, 2001).
4. Of course all documents and objects can also be thought of as only pointing towards the historical event, which itself is always absent. As Barthes suggests, history is a discourse that operates in the context of a subject accessible only through the discourse (1981) – a similar argument is also applicable to performance.

3. Proper Research, Improper Memory

1. As Helen Freshwater observes, the academy 'thrives on the allure of new material' with the requirement to conduct original research boosted by the ability to point to previously unexamined archival documents (2003: 731).
2. Indeed, such is the frequency with which debates surrounding practice-based research in the academy have returned to questions of documentation that Franc Chamberlain has suggested that PaR (Practice as Research) has been in danger of becoming a synonym for DaR (Documentation as Research) (in Thomson 2003: 162). Time and again within this community there is manifest a prominent discourse maintaining the mutual existence and valuation of apparently polar positions of documentation and disappearance (see, for example, Rye 2003 and Thomson 2003 among many others). Practice as Research, therefore, represents one particular environment in which academic attention to documentation has multiplied, with the work often seeking to document (to archive) itself in the same moment that it is created.
3. Of course, perceptions of the value of experiential, live knowledge of performance are equally rooted in quasi-moral assertions, demonstrated not least in the discourses of disappearance explored in Chapter 1.
4. One of the performers Davis identifies as within the canon of feminist theatre is Karen Finley, photographs of whose work are considered in Chapter 7. Indeed, to a great extent it is the powerful and enduring representations manifested in these photographs that has enabled Finley's work to continue to exist within performance history. Without these photographs the work would not be see-able or knowable to academic research.
5. In an article titled 'Performance Remains', Rebecca Schneider presents a description of an alternative kind of performance afterlife, this time located in the bodily or 'flesh memory' of performers and artists (2001), which has similar virtue in terms of its mutability and non-documentness.
6. The uncovering of memory has been a central feature of several audience research projects I have developed. This includes explicit interest in the transformative impact of memory in constructing and assigning meaning to experience, and a conscious development of workshop approaches (primarily drawing and writing based) designed to creatively explore participants' memories of their theatrical experience (see Reason, 2006).
7. There is also an echo here of durational art works, such as Andy Goldsworthy's 'Midsummer Snowballs' (2001), where giant snowballs were placed on city streets, melting and disappearing but leaving a trace in the objects concealed inside.
8. An earlier version of sections of this discussion of performance archives and memory appeared in *New Theatre Quarterly* (Reason, 2003).

4. Self-Representation

1. This chapter explores the documentary attitudes of Forced Entertainment as a company, also examining some examples of academic and archival representations of their work. Later, in Parts II and III, discussion explores examples of the video and photographic representation of Forced Entertainment productions.

2. Unless otherwise referenced, all quotations from Tim Etchells are taken from a personal interview, conducted Sheffield 23 July 2004.
3. Additionally, after this initial function of getting the live productions programmed, promoted and seen, Forced Entertainment's video recordings are made available for sale at inexpensive prices and are now held in numerous libraries around the world, both increasing their accessibility and the reputation of the company.
4. The full text of *A Decade of Forced Entertainment* is reproduced in *Certain Fragments* (1999: 29–46).
5. Another example, and by way of contrast to Forced Entertainment's functional company website, would be *Paradise* (1998 – www.lovebytes.org.uk/paradise), an interactive web project that emerged from the company's model city installation *Ground Plans for Paradise* (1994).
6. Thinking back to Helen Iball's arguments presented in Chapter 3, it is no coincidence that Forced Entertainment, who like DV8 are on the school and undergraduate curriculum, also have a comprehensive web presence and well-maintained archive. They too have taken on some of the responsibilities (and implicitly accepted some of the rewards) that go with achieving canonical status.

5. Video Documents

1. The question of the video recording of theatre is a well-developed theme with an extensive literature, much of which examines the rights and wrong of different approaches to video documentation according to specific purposes and utilitarian functions. This chapter is located within this discourse, including Erenstein (1988), de Marinis (1985), McAuley (1986, 1994), Melzer (1995a, 1995b) and Varney and Fensham (2000). Chapter 6 also draws on the subtly different discourses exploring the relationship between dance and video, such as Dodds (2001), Jordan and Allen (1993) and Mituma (2002).
2. As seen in Chapter 1, discussion of the relationship between the live and non-live in performance has recently been revitalised by Philip Auslander's book *Liveness* (1999). One important theme of this recent debate has been the disruption of binary relationships between the live and non-live, which too easily settle into evaluative comparisons of better and worse, and instead a concentration on the contingent experience of different media (see for example Barker 2003). It is this examination of cultural use and practice, rather than an ontological discussion, which this chapter seeks to pursue in relation to video practices and live performance representation.
3. These chapters focus on a deliberately small number of case studies, with the intention of using them to demark viewing strategies that readers can then apply to their own specific experiences of watching live performance on video. This does mean that some attempts from within the academy to construct alternative, self-reflective and interactive screen documentations have been neglected, including for example Gay McAuley's experimentations with split-screen recordings and Caroline Rye's exploration of approaches to documenting practice-based research. Such specialist representations are inevitably only available to very small audiences and the intention here is to focus on the

video representation of live performance as encountered more widely and thereby having a greater impact on concepts and perceptions of performance.
4. See for example *The Guardian's* 'Virtual Gallery of Lost Art' (Jones, 2002) and also www.artloss.com and www.saztv.com.
5. Unless otherwise referenced, all quotations by Tim Etchells are taken from a personal interview, conducted Sheffield 23 July 2004.
6. Jane Feuer makes a similar argument in relation to television culture, suggesting that 'the "central fact" of television may be that it designed to be watched intermittently, casually and without full concentration' (1983: 15). This experience is increased in relation to video, where there is no perception of a quasi-liveness that can accompany the television broadcast flow.
7. In my use of 'televisual' I am following Catherine Elwers (herself following Stuart Marshall) in using the term to stand for 'the language and representational convention of realism in broadcast television with the implied receptivity if not passivity of the viewer' (2005: 194).
8. And somewhat like the time-stamping of CCTV, several archival institutions and theatre companies overlay their video recordings with spoilers that also act as assertions of their factual status. All National Theatre recordings, for example, carry the legend 'Royal National Theatre Archive' across the top eighth or so of the screen, in part designed to prevent public reuse but also to mark the ideological and aesthetic status of the recording.
9. As Cubitt notes, there is a startlingly powerful cultural competence in media use: 'There is something disturbing about recognising, embedded in our own psyches, the media skills that are as familiar to us as breathing' (1991: 167).

6. Screen Reworkings

1. This is also the case in discussions of film adaptation more widely, such as of novels, where as James Naremore describes there is a history of criticism that is 'inherently respectful of the "precursor text" which is valorised as the original to the screen copy' (2000: 2).
2. Indeed, in recent years the screen-to-stage adaptation of musicals have become commonplace, with examples including *The Lion King* (screen 1994, stage 1997), *The Full Monty* (1997, 2000), *The Producers* (1968, 2001) and *Billy Elliot* (2000, 2005). Other recent non-musical examples include *The Graduate* (1967, 2000), *Festen* (1998, 2003) and *When Harry Met Sally* (1989, 2004).
3. Philip Auslander has recently drawn attention to figures on the number of people who watch or listen to stage theatre or dance performance in non-live media contained within a National Endowment for the Arts 2002 *Survey of Public Participation in the Arts* (published in 2004 and available at www.arts.gov). These statistics suggests that in the United States 22.3% of adults attended the theatre in 2002 while 21% watched theatre in a mediatized form (covering TV, radio, recordings and the internet). Moreover, while live attenders saw an average of 2.3 productions in 2002, non-live viewers saw 6.9 pieces of non-musical theatre. As Auslander writes, 'theatre is being consumed [in the United States] in mediatized forms two to three times more often than it is attended live' (2005: 7). The figures for dance are even more pronounced, with a higher percentage of people watching dance in a mediatized form (13.7%) than live (8.7%). Although it is not entirely clear how this report

defined 'theatre' these figures do suggest that the non-live dissemination of live performance does reach potentially very significant audiences.
4. The only cast change was Cyril Cusack as Uncle Sam, replacing John Normington who was thought too young to play the role on screen – a result of the greater closeness of the camera and the different relationships between realism and the audiences' imagination on stage and screen.
5. André Bazin writes of how deep-focus cinematography allows the camera to take in 'with equal sharpness the whole field of vision' and thereby promotes a respect for 'the continuity of dramatic space and its duration' (in Williams, 1980: 36).
6. Tellingly these arguments are subtly reversed by those who approach this question primarily interested in the camera rather than the stage. Writing about video dance, for example, Dodds argues that the spectator's view of live performance is fixed and static, while in contrast when 'aided by the camera, the spectator is able to see the dance from a multitude of angles and distances' (2001: 31).
7. Viewers' commentary on *The Homecoming* include that 'any evidence of [Ian Holm's] stage work (even filmed) is inestimably valuable' (www.dvdtimes.co.uk). Remarks such as this support the documentary instinct of the AFT to record great and even definitive performances, with this archival value presented as compensation for deficiencies in the film artefacts.
8. With the films distributed simultaneously at a small number of the cinemas, the AFT attempted to operate through a subscription scheme, echoing that employed by many theatres. According to Edie Landau the eventual failure of the project was largely due to problems with the ticketing service provided and a failure of the subscribers to sign up for a third season.
9. The reason for this cannot be that live and non-live dance are more similar than live and non-live drama, but instead is almost certainly because dance (and even more so video dance) is too marginal a form in our culture to have demanded such linguistic distinctions.
10. For these purposes the concepts follow the common and narrow definition in relation to the dominant practices of classical Hollywood realism (for example Bordwell, Staiger and Thompson, 1985).
11. In her analysis of video dance, Dodds echoes similar articulations of the status of video art and its relationship with mainstream forms of screen media (such as Elwers, 2005: 2–4). Indeed, the non-verbal, non-linear and self-reflective aspects of dance on camera directly relate to many characteristics of video art (see also Rush, 2003: 122–3). Similarly, music videos, which frequently make heavy use of dance and movement, have also been described as resistant to dominant screen codes and media (Buckland and Stewart, 1993).
12. Thinking in terms of categorisation, several of DV8's films, including *Enter Achilles* and *The Cost of Living*, have won the award for 'best camera re-work' at the IMZ Dance Screen Festival, rather than being considered under the categories of either 'best screen chorography' or 'best live performance relay'.
13. In a classic formation of the essence of the television screen picture, Herbert Zettl writes 'While in film each frame is actually a static image, the television image is continually moving [...] the scanning beam is constantly trying to complete an always incomplete image. Even if the image on screen seems at

rest, it is structurally in motion. Each television frame is always in a state of becoming. [Live television] lives off the instantaneousness and uncertainty of the moment very much the way we do in actual life. The fact that television can record images and then treat them in a filmic fashion in no way reduces the aesthetic potential and uniqueness of television when used live' (in Feuer, 1983: 13).

14. Much of the language here is drawn from Walter Benjamin's essay 'The Work of Art in the Age of Mechanical Reproduction', which describes the 'aura' of an original work of art as diminishing through reproduction and representation (1970: 211–44). Wurtzler, however, suggests that 'aura', cultural prestige and cult value exist in events determined to be fully live because of, and not despite, the technological encroachment on the very concept of liveness.

7. Photography, Truth and Revelation

1. A shorter version of aspects of this discussion, focusing exclusively on the representation of movement in the photography of Lois Greenfield and Chris Nash, was previously published in *Dance Research Journal* (Reason, 2004).
2. Of course photography does not reproduce the real but 'realism'. The camera is designed to *mechanically* reproduce the dominant idea of representative reality, based upon geometric perspective, with the acceptance of the validity of that reality deeply ingrained. That the camera is a machine, that it is (or rather was) a process of mechanical reproduction, allows an investment of faith in its truthfulness and objectivity that would not be possible with evidently artful or 'manual' depictions of the world. As Chapter 8 explores, this changes in our new age of digital reproduction.
3. All quotations by Graham Brandon are taken from a personal interview, conducted London 19 May 2004.
4. Brandon describes several frustrations resulting from working during photo-calls. Firstly, in terms of his work as an archive photographer, he can only record as much or as little of the performance as is made available to him. However, what he finds personally much more frustrating is the sense that he is only getting a watered down performance for the camera, without the full force, energy or emotional intensity of the actual performance. Instead, photo-calls present short snippets, where the actor is starting cold and rarely able to get fully into the performance. These compromised circumstances, Brandon suggests, are made manifest through the eyes and body language of the performer and inevitably makes itself seen in the resulting photographs.
5. Indeed, it is possible to suggest that more than its revelatory authority it is the reproducibility of the image that is the central aesthetic mode of still photography. This theme is developed in Chapter 8.
6. As with all photojournalism it would be possible to argue that a somewhat exploitative relationship exists behind these images, a voyeuristic quality to cross-cultural records that appear to present as art the suffering of another world. Responses to this possibility depend partly on van Manen's actual relationship with her subjects (knowledge unavailable to us through the photographs) and partly on our personal reading of what is shown in the images.

7. All quotations by Bertien van Manen are taken from personal email correspondence, 6 and 7 July 2004.
8. Reasons for photographers preferring to work in the studio, rather than on the stage, include inadequate lighting levels, the difficulty of knowing what is coming next or of repeating sections for the camera, the angle and distance at which they are required to work, dirty stages or sets that look distracting and fragmented in the background of the photograph and, most importantly, lack of control over the process (Greenfield, 1998: 108; Mitchell, 1999: 73; Nash, 1993: 3).
9. The 'decisive moment' of photography also has links to Sontag's idea of the 'privileged moment' of the selected still image (1979: 18). Merging intriguingly with both these is the tradition in dance photography of attempting to capture the 'perfect moment' of the choreography: the pinnacle of the leap or the perfectly outstretched toe. This is a tradition that is particularly dominant within ballet, where an image that does not capture the perfect moment can simply be dismissed as wrong (Greenfield, 1992: 102; Mitchell, 1999: 74).

8. Photography, Publicity and Representation

1. This is equally true of performance photography in the past, with the camera long being used as a glamorising and promotional tool. For researchers today the promotional photographs of the past become important historical documents, even in instances where they are not literal records of stage performances in the first place (see for example Applebaum, 1976 and Senelick, 1987). This is partly because of the nature of the medium, which always does relate back to compromised but culturally significant perceptions of revelatory authority, as explored in Chapter 7. Equally important, however, is the pragmatic fact that publicity photographs are often the only extant visual record of past performances and performers.
2. All quotations by Andy Catlin are taken from a personal interview, conducted Edinburgh 29 September 2004.
3. All quotations by Euan Myles are taken from a personal interview, conducted Edinburgh 13 July 2004.
4. The exception to this is the use, where possible, of actors from the production. More frequently, however, the play will not be cast far enough in advance, in which case Myles uses stand-ins, operating with an unwritten rule that it is not appropriate to feature a fully identifiable image of someone who is not actually in the production.
5. Unless otherwise referenced, all quotations by Hugo Glendinning are taken from a personal interview, conducted London 18 May 2004.
6. Glendinning's own articulation of this process is as follows: 'In *Club of No Regrets* each of the performers was in a kind of closed of world doing these repeated actions, and the way that they moved was totally isolated, and they'd occasionally run into each other and something would happen, but most of it was mad running around. And so I, after watching a rehearsal, I thought I'll just be with them and see what happens, so I just ran around with them, except I had a camera, that was my job. You know, Richard might be slapping a clock on the wall over and over again and Terry was probably leaning over

the top screaming with a hammer and I can't remember what Claire and Cathy did most of the time, lots of water and flour I think. But I was one of them except I had a camera and I was madly taking pictures.'
7. All quotations by Chris Van der Burght are taken from personal correspondence, letter April 2004.
8. For example, Elisabeth McCausland, writing in the 1930s, declared that: 'The modern dance is like modern painting and modern sculpture, clean, with crisp edges, of rigid and muscular materials. The photographs of the modern dance must be likewise, with great definition of values, not flowing draperies, with forms solidly modelled in light, but not swept by theatrical or stagey spotlights' (In Ewing, 1987: 26).
9. Unless otherwise referenced, all quotations by Chris Nash are taken from a personal interview, conducted London 19 May 2004, and email correspondence, 22 July 2004.
10. Personal interview with Tim Etchells, conducted Sheffield 23 July 2004.
11. There are also several references in Greenfield's writings to how dancers and choreographers have either changed their work or created new work as a result of ideas developed during photographic sessions. The improvisational, creative atmosphere of Greenfield's studio clearly enables movements to be explored freely, with the camera making the new possibilities visible.

9. Reviewing Performance

1. There is a widespread perception that from the second half of the twentieth century onwards reviews have shrunk in length as newspapers have marginalised the arts (see for example Herbert, 1999). In a more accommodating past, Jonathan Kalb claims that George Bernard Shaw could 'spend 1000 words comparing two actresses' complexions' (1993: 167). By such comparison the *India Song* reviews are indeed brief. However, at the upper end of the scale there has not in fact been a significant reduction in length compared with critics such as Edwin Denby in the 1940s or Kenneth Tynan in the 1950s. Whatever the exact or relative length, an important characteristic of all reviewing is that it is easily readable in a single sitting and not an extended discourse.
2. This is not to neglect the perverse joy and wider cultural appeal of critics at their most venomous and vitriolic. Nor indeed, as noted by Diana Rigg in *No Turn Unstoned*, her compilation of 'the worst theatrical reviews', that if critics 'weren't entertaining, they would soon lose their readers' (1982: 21).
3. All quotations by Keith Bruce are taken from a personal interview, conducted Edinburgh 9 August 2004.
4. Certainly, this is the reason why publicity departments provide free press tickets to critics, with reviews performing an important role in the publicising of productions. Indeed the relationship between reviews and marketing is easily blurred, with selective editing of copy and certainly a hint of compromised integrity between some papers and the publicity industry. Michael Billington, for example, relates how he was once introduced at a Disney press junket as 'the critic who didn't like *The Lion King*' (2000) as if such independence of thought was a betrayal of all the good work done by the publicity team.

5. For a satirical take on such possibilities see the exchanges between two theatre critics in Tom Stoppard's play *The Real Inspector Hound* – 'By the way, congratulations, Birdboot ... At the Theatre Royal. Your entire review reproduced in neon! ... You've seen it, of course.' (1993: 10–1).
6. It is possible to argue that reviewing has been untouched by any developments of twentieth-century theory regarding the purpose, nature or ideals of arts criticism – not least in terms of a focus on artistic intention (for example Krauss, 1981: 26). This is not to say that reviewing is criticism practised without thought or theory. Instead, although the review rarely has conscious theory behind it, I would suggest that it possesses a framework of tradition and models of 'common practice' guiding its production.
7. Axiomatically the news media is an ephemeral form – who reads yesterday's papers? However, in practice print endures, becomes archive and becomes history. With performance criticism this process also occurs through the publication of collected reviews in book form.
8. It is not certain how Brustein knows Tynan's verbal descriptions are *exact* equivalents: they may be evocative, theatrical, even inspiring of a sense of live performance, but he cannot know they are exact unless he was there himself.
9. There are also many, possible apocryphal, stories of critics taking advantage of the opportunity to write such contextualising sections of their reviews in advance of seeing a performance. (And thereby bypassing some of the time constraints involving in writing to newspaper deadlines.)
10. As a character in Will Stillman's film *Metropolitan* (1990) puts it, 'I don't read novels. I prefer good literary criticism – that way you get both the novelist's ideas and the critic's thinking'.

10. Writing the Live

1. The description of the olfactory experience is a particularly interesting one here, for smell is often described as something powerfully rooted in a particular time and space. And indeed, something particularly rooted in the live experience (see particularly Banes, 2001).
2. Indeed, Keith Bruce of *The Herald* notes how reviews typically fit an almost predetermined structure of paragraphs, with what goes in each one becoming ingrained to the extent that the content can begin to write itself.
3. It is also, Beckerman points out, a technique essentially borrowed from literary criticism, which he sees as excusable to a certain extent as an established method of dealing with material. However, as he argues, 'the habit of mind that chooses to treat a play as a collection of strands inhibits an appreciation of it as a sequence of total experiences' (1979: 36).
4. Similarities can also be drawn with Stanley Fish's exploration of how the experience of the reader of literature is in response to a 'temporal flow', not the 'whole utterance', and his subsequent attempts in reader response criticism to devise a method that reflected this nature of the experience (1980).
5. While it is clearly possible for collective audiences of non-live performances to respond communally, the live performance audience also possesses a communal influence on the act of creation. This reciprocity, the ability not

only for the performer to influence the audience but also for the audience to influence the performance, produces a mutually manipulative, empathetic relationship between the audience and performer that can exist only during live performances.

6. A succinct example of this awareness of the embodiment of others, our awareness of what might be seen as meta-physicality, can be drawn from the everyday. Lifts, particularly, are contemporary sites where the physical presence of others is experienced without actual physical contact. The confined space, lack of air and a sense of awkwardness all encourage greater awareness of our surroundings, producing a sense of proximity with the lives of others. A similar process occurs in the live performance venue, where various qualities of space, occasion and atmosphere all encourage a greater sense of awareness. Indeed, specialised performance spaces are designed to enable and enhance these qualities of empathy and intersubjectivity. When redesigning the Royal Court Theatre, reopened in February 2000, theatre consultant Maxwell Hutchison was alert to these aspects, seeking to use not only eye to eye but also thigh to thigh contact between the audience. The inclusion of leather seats, a luxurious but important addition, was intended to further encourage bodily and sensual awareness.

7. Both British theatre critics, Kenneth Tynan's principal period of writing extends from 1944 to 1963; Michael Billington's collected reviews covers 1972 to 1994 and he continues to write today. Tynan dominates discussions on post-war theatre reviewing; while Billington has become one of the best-known British drama critics of the last 30 years. Taken together their work displays a number of similarities and a range of mirrored themes and shared techniques.

8. Billington's criticism is subject to sustained scrutiny by Susan Melrose in an essay that seeks to explore the limitations and conservative nature of his engagement with performance, and especially with Live Art. In particular Melrose's identifies a 'taste for nameable categories' or a need to '*know what he knows*' and like what fits in with his conception of theatrical form and value (1994: 74–8). Additionally, Melrose criticises Billington's attachment (or perhaps even need) to construct empathetic and sympathetic relationships with believable characters in believable fictions, although in doing so she is perhaps also criticising the particular kind of attachment that produces the engaged and evocative descriptions of actors' bodies that are found in the writings of both Billington and Tynan.

9. For example, Franz Kafka describes how the experience of reading effects him physically and bodily: 'I read sentences of Goethe as though my whole body were running down the stresses' (in Burnshaw, 1970: 268).

10. Porter's use of such synaesthetic language is his subtle response to the problems of writing about music, of describing the quality of an orchestra's sound. This is a problem that he is clearly aware of, occasionally discussing the difficulties explicitly, asking in one review 'how does one find words to relate adventures of the spirit?' (1979: 427). In another review, Porter observes that 'It is easier to declare a performance ineffable than to find words to describe it' (1979: 599). Although he never discusses possible answers to such questions, his use of synaesthetic language certainly points the way for an effective and embodied live performance writing.

Performance Works Cited

This bibliography lists the main performance works cited in this book. Relevant websites are provided where possible (accessed January 2006).

Adventures in Motion Pictures
Swan Lake (1995), Choreography Matthew Bourne, Music P.I. Tchaikovsky.

American Theater Lab
American Theater Lab Concert (1979), Choreography Aileen Passloff and Toby Armour, Music played by Richard Busch.

Penny Arcade
While You Were Out (1985).

Companyia Carles Santos, www.carles-santos.com
Ricardo i Elena (2000), Script, music and direction Carles Santos.

Complicite, www.complicite.org
The Chairs (1997), Direction Simon McBurney, Written by Eugène Ionesco.

Damaged Goods, www.damagedgoods.be
appetite (1998), Choreography Meg Stuart, with Ann Hamilton.
Visitors Only (2003), Choreography Meg Stuart, with Anna Viebrock.

DV8, www.dv8.co.uk
Dead Dreams of Monochrome Men (1988), Artistic Direction Lloyd Newson, Music Sally Herbert.
Filmed 1989. Director David Hinton. Dance Videos and LWT.
Strange Fish (1992), Direction Lloyd Newson, Music Jocelyn Pook (vocal) and Adrian Johnston (soundtrack).
Filmed 1992. Direction David Hinton. Dance Videos and BBC.
Enter Achilles (1995), Direction Lloyd Newson, Music Adrian Johnston.
Filmed 1995. Direction Clara van Gool. Dance Videos and BBC.
Cost of Living (2003), Direction Lloyd Newson, Music Paul Charlier.
Filmed 2004. Direction Lloyd Newson. DV8 Films and Channel 4 Television.

The Featherstonehaughs, www.thecholmondeleys.org
Immaculate Conception (1992), Choreography Lea Anderson.
The Featherstonehaughs Draw on the Sketchbooks of Egon Schiele (1997), Choreography Lea Anderson, Music D. Madden.

Karen Finley, www.karenfinley.net
I'm an Ass Man (Club Show) (1987)

Forced Entertainment, www.forced.co.uk
200% and Bloody Thirsty (1987), Artistic Direction Tim Etchells
Video document recorded Nottingham 1998.

    Some Confusions in the Law About Love (1989), Artistic Direction Tim Etchells.
      Video documentation recorded Sheffield 1991.
    Marian & Lee (1991), Artistic Direction Tim Etchells.
      Video documentation recorded Sheffield 1991.
    Emanuelle Enchanted (1992), Artistic Direction Tim Etchells.
      Photographic project *Cardboard Sign Photographs* (1992),
      Photography Hugo Glendinning.
    Club of No Regrets (1993), Artistic Direction Tim Etchells.
      Video documentation recorded London 1993.
    A Decade of Forced Entertainment (1994), Artistic Direction Tim Etchells.
    Nights in This City (1995 and 1997), Artistic Direction Tim Etchells.
    First Night (2001), Artistic Direction Tim Etchells.

Het Zuidelijk Toneel
    India Song (1998), Direction Ivo van Hove, Music Henry de Wit, Written by Marguerite Duras.

The Old Vic
    Henry IV (1946, revival), Direction John Burrell, Written by William Shakespeare.

Ro Theatre, www.rotheater.nl
    Nachtasiel [*The Lower Depths*] (1998), Direction Alize Zandwijk, Written by Maxim Gorky.

Royal Shakespeare Company, www.rsc.org.uk
    The Homecoming (1965), Direction Peter Hall, Written by Harold Pinter.
      Filmed 1973. Direction Peter Hall. American Film Theatre.
    Uncle Vanya (1974), Direction Nicol Williamson, Written by Anton Chekhov.

Annie Sprinkle, www.anniesprinkle.org
    Post Porn Modernist (1990).

Traverse Theatre Company, www.traverse.co.uk
    Olga (2001), Direction Lynne Parker, Written by Laura Ruohonen.
    Outlying Islands (2002), Direction Philip Howard, Written by David Greig.
    When the Bulbul Stopped Singing (2004), Direction Roxana Silbert, Written by Raja Shehadeh.

Trisha Brown Company
    Line Up, including 'Spanish Dance' (1977), Choreography Trisha Brown.

Ultima Vez, www.ultimavez.com
    Scratching the Inner Fields (2001), Choreography Wim Vandekeybus, Music Eavesdropper.

Yolande Snaith Theatre Dance
    Blind Faith (1997), Choreography Yolande Snaith.

Bibliography

Abercrombie, Nicholas, and Brian Longhurst. *Audiences: A Sociological Theory of Performance and Imagination.* (London: Sage Press, 1998).
Advertising Standards Agency. *Phoenix Dance Theatre.* 25 June 2003. Available: www.asa.org.uk/adjudications/. Accessed: September 2004.
Allen, Dave. 'Screening Dance.' *Parallel Lines: Media Representations of Dance.* Eds Stephanie Jordan and Dave Allen. (London: John Libbey, 1993). 1–35.
Aloff, Mindy. *It's Not Ephemera after All.* 2001. National Initiative to Preserve America's Dance. Available: www.danceusa.org/NIPAD/nextsteps7.html. Accessed: September 2001.
Alvarado, Manuel. 'Photographs and Narrativity.' *Representation and Photography.* Eds Manuel Alvarado, Edward Buscombe and Richard Collins. (London: Palgrave, 2001). 148–63.
Anonymous. 'The Russian Ballet.' *T.P.'s Weekly.* (London: December, 1913). 764.
Appelbaum, Stanley, Ed. *The New York Stage: Famous Productions in Photographs.* (New York: Dover Publications, 1976).
Artaud, Antonin. *Theatre and its Double.* 1938. Trans. M.C. Richards. (New York: Grove Press, 1958).
Arts Archive. 2005. Arts Documentation Unit. Available: www.arts-archives.org. Accessed: September 2005.
Auslander, Philip. 'No-Shows: The Head Count from the NEA.' *The Drama Review* 49.1 (2005). 5–9.
——. *Liveness: Performance in a Mediatized Culture.* (London: Routledge, 1999).
Banes, Sally. 'Olfactory Performances.' *The Drama Review* 45 (2001). 68–76.
——. *Writing Dancing in the Age of Postmodernism.* (Hanover: Wesleyan University Press, 1994).
Barba, Eugenio. 'Efermaele: "That Which Will Be Said Afterwards".' *The Drama Review* 36.2 (1992). 77–80.
——. 'Four Spectators.' *The Drama Review* 34.1 (1990). 96–100.
Barker, Martin. '*Crash*, Theatre Audiences, and the Idea of "Liveness".' *Studies in Theatre and Performance* 23.1 (2003). 21–39.
Barthes, Roland. *Camera Lucida: Reflections on Photography.* Trans. Richard Howard. (London: Flamingo, 1984).
——. 'The Discourse of History.' *Comparative Criticism* 3 (1981). 7–20.
——. 'The Grain of the Voice.' *The Responsibility of Forms.* 1977. Trans. Richard Howard. (London: Basil Blackwell, 1985). 267–77.
Baudrillard, Jean. *Simulations.* Trans. Paul Foss, Paul Patton and Philip Beitchman. (New York: Semiotext[e], 1983).
Bazin, André. *What is Cinema?* Trans. Hugh Gray. (Berkeley: University of California Press, 1967).
Beckerman, Bernard. *Dynamics of Drama.* (New York: Drama Book Specialists, 1979).
Beloff, Hella. *Camera Culture.* (London: Basil Blackwell, 1985).

Benjamin, Walter. 'The Work of Art in the Age of Mechanical Reproduction.' Ed. Hannah Arendt. Trans. Harry Zohn. 1936. *Illuminations*. (London: Jonathan Cape, 1970). 211–44.

Berger, John. *Ways of Seeing*. (London: BBC/Penguin, 1972).

Billington, Michael. 'We Will Not Be Muzzled.' Comment. *The Guardian* (28 June 2000) sec. G2: 14.

——. *One Night Stands: A Critic's View of Modern British Theatre*. (London: Nick Hern Books, 1994).

Billman, Larry. 'Music Video as Short Form Dance Film.' *Envisioning Dance on Film and Video*. Eds Judith Mituma, Elizabeth Zimmer and Dale Ann Stieber. (New York: Routledge, 2002). 12–20.

Billy Rose Theatre Collection. 2005. New York Public Library. Available: www.nypl.org/research/lpa/the/the.html. Accessed: September 2005.

Blau, Herbert. *The Audience*. (Baltimore, MD: Johns Hopkins University Press, 1990).

——. *Take Up the Bodies: Theatre at the Vanishing Point*. (Urbana, IL: University of Illinois Press, 1982).

Booth, John E. *The Critic, Power, and the Performing Arts*. (New York: Columbia University Press, 1991).

Bordwell, David, Janet Staiger and Kristin Thompson. *The Classical Hollywood Cinema: Film Style and Mode of Production to 1960*. (London: Routledge, 1985).

Bradley, Harriet. 'The Seductions of the Archive: Voices Lost and Found.' *History of the Human Sciences* 12.2 (1999). 107–22.

Breslauer, Jan. 'Man with a Press Camera.' *Theater* 18.2 (1987). 34–6.

Brooks, Bonnie. *Use It or Lose It: Artists Keep Dance Alive*. 2001. National Initiative to Preserve America's Dance. Available: www.danceusa.org/NIPAD/ nextsteps1.html. Accessed: September 2001.

Brooks, Virginia. 'From Méliès to Streaming Video: A Century of Moving Dance Images.' *Envisioning Dance on Film and Video*. Eds Judy Mituma, Elizabeth Zimmer and Dale Ann Stieber. (New York: Routledge, 2002). 54–60.

Brown, Richard Harvey and Beth Davis-Brown. 'The Makings of Memory: The Politics of Archive, Libraries and Museums in the Construction of National Consciousness.' *History of the Human Sciences* 11.4 (1998). 17–32.

Buckland, Theresa Jill and Elizabeth Stewart. 'Dance and Music Video.' *Parallel Lines: Media Representations of Dance*. Eds Stephanie Jordan and Dave Allen. (London: John Libbey, 1993). 51–79.

Burgin, Victor. 'Looking at Photographs.' *Representation and Photography*. Eds Manuel Alvarado, Edward Buscombe and Richard Collins. (London: Palgrave, 2001). 65–75.

Burnshaw, Stanley. *The Seamless Web: Language-Thinking, Creature-Knowledge, Art-Experience*. (London: Allen Lane, 1970).

Burrows, David. 'A Dynamical Systems Perspective on Music.' *Journal of Musicology* 15.4 (1997). 529–45.

——. *Sound, Speech and Music*. (Amherst: University of Massachusetts Press, 1990).

——. 'Music and the Biology of Time.' *Perspectives on New Music* 11.2 (1972). 241–8.

Carr, C. 'The Karen Finley Makeover: A Persecuted Artist Gets Past Her Suffering.' *Village Voice* (8–14 November 2000) Available: www.villagevoice.com/news/ 0045, carr, 196371.html Accessed: May 2006.

Cass, Joan. 'The Uses of Criticism.' *Dance Experience: Readings in Dance Appreciation*. Eds Myron Nadel and Constance Nadel. (New York: Praeger Publishers, 1970).

Christiansen, Rupert. 'Silence Is Not Golden.' Review. *The Daily Telegraph* (2 September 1999).

Clifton, Thomas. *Music as Heard: A Study in Applied Phenomenology*. (New Haven: Yale University Press, 1983).

Clurman, Harold. *The Collected Works of Harold Clurman*. Eds Marjorie Loggia and Glenn Young. (New York: Applause Books, 1994).

Connor, Steven. *Postmodernist Culture*. (Oxford: Basil Blackwell, 1989).

Cooper, Neil. *India Song*. Review. *The Times* (2 September 1999) Features.

Copeland, Roger. 'Between Description and Deconstruction.' *Routledge Dance Studies Reader*. Ed. Alexander Carter. (London: Routledge, 1998). 98–107.

——. 'The Presence of Mediation.' *The Drama Review* 34.4 (1990). 28–44.

Cousin, Geraldine. *Recording Women: A Documentation of Six Theatre Productions*. Contemporary Theatre Series. Ed. Franc Chamberlain. (Amsterdam: Harwood Academic Publishers, 2000).

Cowley, Malcolm. *Writers at Work: The Paris Review Interviews*. (London: Mercury Books, 1962).

Croce, Arlene. *Afterimages*. (London: Adam & Charles Black, 1978).

Cubitt, Sean. *Timeshift: On Video Culture*. (London: Routledge, 1991).

Cunningham, Merce. *Changes: Notes on Choreography*. Ed. Frances Starr. (New York: Something Else Press, 1968).

Cutler, Anna. 'Abstract Body Language: Documenting Women's Bodies in Theatre.' *New Theatre Quarterly* 14.2 (1998). 111–18.

Dance Heritage Coalition. 2001. Available: www.danceheritage.org. Accessed: September 2001.

Daniel, Claire. 'Archival Sound and Video Recordings of Live Performance.' Unpublished MA Thesis. University College, London, 2004.

Davidson, Hugh. 'The Critical Position of Roland Barthes.' *Criticism: Speculative and Analytical Essays*. Ed. LS Dembo. (Madison, London: University of Wisconsin Press, 1968).

Davis, Jill. 'Goodnight Ladies. On the Explicit Body in Performance.' *New Theatre Quarterly* 15.2 (1999). 183–7.

de Certeau, Michel. *The Writing of History*. 1975. Trans. Tom Conley. (New York: Columbia University Press, 1988).

Deleuze, Gilles. *Difference and Repetition*. Trans. Paul Patton. (London: Athlone, 1994).

de Marinis, Marco. 'A Faithful Betrayal of Performance: Note on the Use of Video and Theatre.' *New Theatre Quarterly* 1.4 (1985). 383–9.

Denby, Edwin. *Dance Writings*. Ed. Robert Cornfield and William MacKay. (New York: Alfred A Knopf, 1986).

Derrida, Jacques. *Archive Fever: A Freudian Impression*. Trans. Eric Prenowitz. (Chicago: University of Chicago Press, 1995).

Dodds, Sherril. *Dance on Screen: Genres and Media from Hollywood to Experimental Art*. (London: Palgrave, 2001).

Durant, Alan. *Conditions of Music*. (London: Macmillan, 1984).

Early, Fergus, Ed. *The Body Electric: An Exhibition of Ballet and Dance Photography from 1859 to the Present Day*. (York: The Impressions Gallery of Photography, 1984).

Elam, Keir. *The Semiotics of Theatre and Drama*. (London: Methuen, 1980).
Elsom, John. *Post-War British Theatre Criticism*. (London: Routledge and Kegan Paul, 1981).
Elwers, Catherine. *Video Art: A Guided Tour*. (London: I. B. Travis and Co., 2005).
Erenstein, Robert L. 'Theatre Iconography: An Introduction.' *Theatre Research International* 22.3 (1997). 185–9.
———. Ed. *Theatre and Television*. (Amsterdam: International Theatre Bookshop, 1988).
Etchells, Tim. *Certain Fragments: Contemporary Performance and Forced Entertainment*. (London: Routledge, 1999).
Etchells, Tim, and Richard Lowdon. '*Emanuelle Enchanted* (or a Description of This World as if it Were a Beautiful Place): Notes and Documents.' *Contemporary Theatre Review* 2.2 (1994). 9–24.
Evans, Martyn. *Listening to Music*. (London: Macmillan Press, 1990).
Evans, Richard. *In Defence of History*. (London: Granta Books, 1997).
Ewing, William A. *The Fugitive Gesture: Masterpieces of Dance Photography*. (London: Thames and Hudson, 1987).
Farquharson, Alex, Ed. *Citibank Photography Prize 2003*. (London: The Photographers' Gallery, 2003).
Feingold, Michael. 'Maybe, in Our Hemmed-in World, One Step Back Could Mean Two Steps Forward.' Review. *The Village Voice* (6 September 2000).
Féral, Josette. 'Performance and Theatricality: The Subject Demystified.' *Modern Drama* 25.1 (1982). 170–81.
———. ' "The Artwork Judges Them": The Theatre Critic in a Changing Landscape.' *New Theatre Quarterly* 16.4 (2000). 307–14.
Feuer, Jane. 'The Concept of Live Television: Ontology as Ideology.' *Regarding Television: Critical Approaches – an Anthology*. Ed. E. Ann Kaplan. (Los Angeles: American Film Institute, 1983).
Fish, Stanley. *Is There a Text in This Class? The Authority of Interpretive Communities*. (Cambridge, Mass: Harvard University Press, 1980).
Fisher, Mark. *India Song*. Review. *The Herald* (1 September 1999). 5.
Foucault, Michel. *The Archaeology of Knowledge*. Trans. A M Sheridan Smith. (London: Tavistock Books, 1972 [1969]).
Freebalm, Alison. *India Song*. Review. *The Stage* (9 September 1999).
Freshwater, Helen. 'The Allure of the Archive.' *Poetics Today* 24.4 (2003). 729–58.
Fry, Christopher. *An Experience of Critics*. (London: Perpetua, 1952).
Fuchs, Elinor. 'Presence and the Revenge of Writing: Re-Thinking Theatre after Derrida.' *Performing Arts Journal* 26.7 (1985). 163–73.
Garner, Stanton B. *Bodied Spaces: Phenomenology and Performance in Contemporary Drama*. (Ithaca, NY: Cornell University Press, 1994).
Giesekam, Greg. 'A View from the Edge.' *Contemporary Theatre Review* 2.2 (1994). 115–29.
Goldberg, RoseLee. *Performance Art: From Futurism to the Present*. 1979. (London: Thames and Hudson, 1988).
———. *Performance Live Art since the '60s*. (London: Thames and Hudson, 1998).
Greenfield, Lois. *Airborne: The New Dance Photography of Lois Greenfield*. (London: Thames and Hudson, 1998).
———. *Breaking Bounds: The Dance Photography of Lois Greenfield*. Ed. William A Ewing. (London: Thames and Hudson, 1992).

——. *www.loisgreenfield.com*. 2004. Lois Greenfield Studios. Available: www.lois.Greenfield.com/about/index.html. Accessed: September 2004.
Guest, Ann Hutchinson. *Dance Notation: The Process of Recording Movement on Paper*. (London: Dance Books, 1984).
Hall, Fernau. 'Dance Notation and Choreology.' *What is Dance? Readings in Theory and Criticism*. Ed. Roger Copeland and Marshall Cohen. (Oxford: Oxford University Press, 1983). 390–9.
Hall, Peter. '*The Homecoming*: Interview with Peter Hall.' American Film Theatre Collection. DVD. Ed. Peter Hall: InD, 2004.
Harper, Douglas. 'Reimagining Visual Methods: Galileo to *Neuromancer*.' *Handbook of Qualitative Research*. Eds N.K. Denzin and Y.S. Lincoln. (Thousand Oaks, CA: Sage, 2000). 717–32.
Heathfield, Adrian. 'End Time Now.' *Small Acts: Performance, the Millennium and the Marking of Time*. Ed. Adrian Heathfield. (London: Black Dog, 2000). 104–11.
Heathfield, Adrian, and Andrew Quick. 'Editorial: On Memory.' *Performance Research* 5.3 (2000). 1–2.
Helmer, Judith, and Florian Malzacher. *Not Even a Game Anymore: The Theatre of Forced Entertainment*. (Berlin: Alexander, 2004).
Herbert, Ian. 'Writing in the Dark: Fifty Years of British Theatre Criticism.' *New Theatre Quarterly* 15.3 (1999). 236–46.
Hodges, Nicola. 'Videos on Performance Art or Live Art.' *Art and Design: Performance Art into the 90s*. 38 (1994). xvi.
Hoggard, Liz. 'A Star is Bourne.' *The Observer* (21 September 2003). sec. Review: 5.
Husserl, Edmund. *Cartesian Meditations: An Introduction to Phenomenology*. 1933. Trans. Dorian Cairns. (The Hague: Martinus Nijholt, 1960).
Iball, Helen. 'Dusting Ourselves Down.' *Performance Research* 7.4 (2002). 59–63.
Imhauser, Marcelle. 'Are the Dice Loaded?' *Theatre and Television*. Ed. Robert L. Erenstein. (Amsterdam: International Theatre Bookshop, 1988). 97–100.
Jacobson, Colin, and Mark Haworth-Booth. 'Making Magic.' *Reportage* Spring 4 (1994). 32–3.
Jarrell, Randall. *Poetry and the Age*. (London: Faber and Faber, 1960).
Jenkins, Keith. *Why History? Ethics and Postmodernity*. (London: Routledge, 1999).
Jerome Robbins Dance Division. 2005. New York Public Library. Available: www.nypl.org/research/lpa/dan/danabout.html. Accessed: September 2005.
Johnson, Catherine, and Allegra Fuller Snyder. *Securing Our Dance Heritage: Issues in the Documentation and Preservation of Dance*. (Washington DC: Council on Library and Information Resources, 1999.
Jones, Amelia. ' "Presence" in Absentia: Experiencing Performance as Documentation.' *Art Journal* 56.4 (1997).
Jones, Jonathan. *Missing Masterpieces: Virtual Gallery of Lost Art*. 2002. *The Guardian*. Available: www.guardian.co.uk/arts/arttheft/page/0,13883,1034155,00.html. Accessed: June 2005.
Jonson, Ben. *Ben Jonson: The Complete Masques*. 1608. Ed. Stephen Orgel. (New Haven: Yale University Press, 1969).
Jordan, Stephanie, and Dave Allen, Eds *Parallel Lines: Media Representations of Dance*. (London: John Libbey, 1993).
Jowitt, Deborah. *Dance Beat: Selected Views and Reviews 1967–76*. (New York: Marcel Dekker, 1977).

Kalb, Jonathan. *Free Admissions: Collected Theater Writings*. (New York: Limelight Editions, 1993).
Kalisz, Richard. 'Televized Theatricality or Televized Theatre?' *Theatre and Television*. Ed. Robert L. Erenstein. (Amsterdam: International Theatre Bookshop, 1988). 79–82.
Kaye, Nick. *Site-Specific Art: Performance, Place and Documentation*. (London: Routledge, 2000).
——. 'Live Art: Definition and Documentation.' *Contemporary Theatre Review* 2.2 (1994). 1–7.
Keller, Hans. *Essays on Music*. Ed. Christopher Wintle. (Cambridge: Cambridge University Press, 1994).
——. *Criticism*. Ed. Julian Hogg. (London: Faber and Faber, 1987).
Kirby, Michael. 'Criticism: Four Faults.' *The Drama Review* 18.3 (1974a). 59–68.
——. Ed. *The New Theatre: Performance Documentation*. (New York: New York University Press, 1974b).
Kirkley, Richard Bruce. 'Image and Imagination: The Concept of Electronic Theatre.' *Canadian Theatre Review* 64. Fall (1990). 4–12.
Koestler, Arthur. *Insight and Outlook*. (London: Macmillan, 1949).
Krauss, Rosalind. 'The Effects of Critical Theories on Practical Criticism, Cultural Journalism and Reviewing.' *Partisan Review* 48.1 (1981). 26–35.
Krummel, D.W. 'The Memory of Sound: Observations on the History of Music on Paper.' Engelhard Lecture on the Book. (Washington, DC: Library of Congress, 1987).
Landau, Edie. '*The Maids*: Interview with Edie Landau.' American Film Theatre Collection. DVD. Ed. George Scott. Dir. Christopher Miles: InD, 2004.
Langer, Susanne K. *Feeling and Form: A Theory of Art*. (London: Routledge and Kegan Paul, 1953).
Lawson, Mark. 'Front Row.' London: BBC Radio 4, 19 July, 2004.
Leclercq, Nicole. *Le Photo de Théâtre: Reportage? Création?* 2001. International Association of Libraries and Museums of the Performing Arts. Available: http://www.theatrelibrary.org/sibmas/congresses/sibmas96/hels06.html. Accessed: October 2005.
Live Art Archive. 2000. Nottingham Trent University. Available: http://art.ntu.ac.uk/liveart/newnext.htm. Accessed: September 2005.
Lockyer, Bob. 'A Home for Our Heritage.' *Dancing Times* October (2000). 41.
Macaulay, Alastair. Ed. *Matthew Bourne and His Adventures in Motion Pictures*. (London: Faber and Faber, 2000).
——. 'Duras's Play Dissected to Dreadful Effect.' Review. *Financial Times* (2 September 1999).
Mackintosh, Iain. *Architecture, Actor and Audience*. (London: Routledge, 1993).
Marien, Mary Warner. *Photography and its Critics: A Cultural History 1839–1900*. (Cambridge: Cambridge University Press, 1997).
Marigny, Chris de, and Barbara Newman. 'Progressive Programming: Interviews with Michael Kustow.' *Parallel Lines: Media Representations of Dance*. Eds Stephanie Jordan and Dave Allen. (London: John Libbey, 1993). 81–98.
McAdams, Dona Ann. *Caught in the Act: A Look at Contemporary Multi-Media Performance*. (New York: Aperture, 1996).
McAuley, Gay, Ed. *The Documentation and Notation of Theatrical Performance*. (Sydney: Sydney Association for Studies in Society and Culture, 1986).

——. 'The Video Documentation of Theatrical Performance.' *New Theatre Quarterly* 10.38 (1994). 183–94.
McMillan, Joyce. *India Song*. Review. *The Scotsman* (1 September 1999). 13.
McQuire, Scott. *Visions of Modernity: Representation, Memory, Time and Space in the Age of the Camera*. (London: Sage, 1988).
Melrose, Susan. 'Please, Please Me: "Empathy" and "Sympathy" in Critical Metapraxis.' *Contemporary Theatre Review* 2.2 (1994). 73–83.
Melzer, Annabelle. ' "Best Betrayal": The Documentation of Performance on Film and Video, Part 1.' *New Theatre Quarterly* 11.42 (1995a). 147–57.
——. ' "Best Betrayal": The Documentation of Performance on Video and Film, Part 2.' *New Theatre Quarterly* 11.43 (1995b). 259–76.
Merleau-Ponty, Maurice. *Phenomenology of Perception*. Trans. Colin Smith. (London: Routledge and Kegan Paul, 1962).
Michelson, Sarah. 'An Interview with Deborah Jowitt.' *Movement Research Journal* 25.Dance Writing (2002). Available: http://movementresearch.org/performancejournal/pj25/25index.html Accessed: May 2006.
Miller, Jonathan. *Subsequent Performances*. (London: Faber and Faber, 1986).
Mitchell, Jack. 'Capturing Emotion in Motion.' *Dance Magazine* December (1999). 66–75.
Mituma, Judith, Elizabeth Zimmer, and Dale Ann Stieber. Eds *Envisioning Dance on Film and Video*. (New York: Routledge, 2002).
Morris, Peter, and Peter Hampson. *Imagery and Consciousness*. (London: Academic, 1983).
Naremore, James. Ed. *Film Adaptation*. (London: Athlone Press, 2000).
Nash, Chris. *A Glance at the Toes: The Dance Photography of Chris Nash*. Contemporary Portfolio Series. (Croydon, Surrey: Creative Monochrome, 1993).
——. *Stop Motion*. (London: Fiat Lux, 2001).
NIPAD. 2001. National Initiative to Preserve America's Dance. Available: http://www.danceusa.org/NIPAD/nipad.html. Accessed: September 2001.
Otake, Eiko. 'A Dancer Behind the Lens.' *Envisioning Dance on Film and Video*. Eds Judith Mituma, Elizabeth Zimmer and Dale Ann Stieber. (New York: Routledge, 2002). 82–8.
Parry, Jann. 'Now You See Them.' *The Observer* (29 May 2005) sec. Review: 11.
Pavis, Patrice. *Dictionary of the Theatre: Terms, Concepts and Analysis*. Trans. Christine Shantz. (Toronto: University of Toronto Press, 1998).
——. *Theatre at the Crossroads of Culture*. Trans. Loren Kruger. (London: Routledge, 1992).
——. *Languages of the Stage: Essays in the Semiology of Theatre*. (New York: Performing Arts Journal Publications, 1982).
Pearson, Joanne. 'Stilling Bodies/Animating Texts: Isadora Duncan and the Archive.' *Performance Research* 7.4 (2002). 108–15.
Peña, Richard. '*The Homecoming*: Interview with Richard Peña.' American Film Theatre Collection. DVD. Dir. Peter Hall: InD, 2004.
Phelan, Peggy. *Unmarked: The Politics of Performance*. (London: Routledge, 1993).
Phelan, Peggy. 'Performance, Live Culture and Things of the Heart' (In conversation with Marquard Smith). *Journal of Visual Culture* 2.3 (2003). 291–302.
Phelan, Peggy, and Jill Lane. Eds *The Ends of Performance*. (New York: New York University Press, 1998).
Phillips, Sîan. *Public Places*. (London: Hodder and Stoughton, 2001).

Pollock, Della. 'Performing Writing.' *The Ends of Performance*. Eds Peggy Phelan and Jill Lane. (New York: New York University Press, 1998). 73–103.

Pontbriand, Chantal. 'The Eye Finds No Fixed Point.' *Modern Drama* 25.1 (1982). 154–61.

Porter, Andrew. *Music of Three Seasons 1974–7*. (London: Chatto and Windus, 1979).

——. *Musical Events a Chronical 1980–3*. (London: Grafton Books, 1988).

Potter, Michelle. 'A National Archive.' *Dancing Times* November (2000). 120.

Reason, Matthew. 'Archive or Memory? The Detritus of Live Performance.' *New Theatre Quarterly* 19.1 (2003). 82–9.

——. 'Still Moving: The Revelation or Representation of Dance in Still Photography.' *Dance Research Journal* 35/36.2/1 (2004). 43–67.

——. 'Young Audience and Live Theatre, Part 1: Methods, Participation and Memory in Audience Research.' *Studies in Theatre and Performance* 26.2 (2006).

——. 'Young Audience and Live Theatre, Part 2: Perceptions of Liveness in Performance.' *Studies in Theatre and Performance* 26.3 (2006).

Rebellato, Dan. 'Playwrighting and Globalization: Towards a Site-Unspecific Theatre.' *Contemporary Theatre Review*. 16.1 (2006). 97–113.

Rigg, Diana, Ed. *No Turn Unstoned: The Worst Ever Theatrical Reviews*. (London: Arrow, 1982).

Rogoff, Gordon. 'Theatre Criticism: The Elusive Object, the Fading Craft.' *Performing Arts Journal* 26/27 (1985). 133–41.

Rosenberg, Douglas. *Dancing for the Camera*. 2001. Available: www.dvpg.net/docs/adfessay.pdf. Accessed: October 2005.

——. *Video Space: A Site for Choreography*. 2002. Dziga Vertov Performance Group. Available: www.dvpg.net/docs/videospace.pdf. Accessed: October 2005.

Rosenthanl, Daniel. *Shakespeare on Screen*. (London: Hamlyn, 2000).

Royal Shakespeare Company. *The Essential Shakespeare – Live*. 2005. Press Release. Available: www.rsc.org.uk/press/420_3138.aspx. Accessed: October 2005.

Rush, Michael. *Video Art*. (London: Thames and Hudson, 2003).

Russell, Susan. 'Corporate Theatre: The Revolution of the Species.' Unpublihsed MA Thesis. Florida State, 2003. Available: http://etd.lib.fsu.edu/theses/available/etd-11172003-215153/ Accessed: May 2006.

Rye, Caroline. *Video Writing: The Documentation Trap, or the Role of Documentation in the Practice as Research Debate*. 2003. PARIP. Available: www.bris.ac.uk/parip/webpaper_rye.doc. Accessed: October 2005.

Salter, Alan. *Perspectives on Notation. Vol. 1 and 2. Notation and Dance*. (London: Laban Centre for Notation and Dance, University of London Goldsmiths' College, 1978).

Sartre, Jean-Paul. 'The Author, the Play and the Audience.' *Sartre on Theatre*. *L'Express* 17 September 1959. Eds M. Contat and M. Rybalka. Trans. F. Jellinek. (London: Quartet Books, 1976).

Savedoff, Barbara E. *Transforming Images: How Photography Complicates the Picture*. (Ithaca: Cornell University Press, 2000).

Schechter, Joel. 'Theater and Photography Editorial.' *Theater* 18.2 (1987). 4.

Schneider, Rebecca. 'Performance Remains.' *Performance Research* 6.2 (2001). 100–8.

Searle, Judith. 'Four Drama Critics.' *The Drama Review* 18.3 (1974). 5–23.

Senelick, Laurence. 'Early Photographic Attempts to Record Performance Sequence.' *Theatre Research International* 22.3 (1997). 255–64.

——. 'Melodramatic Gesture in Carte-De-Visite Photographs.' *Theater* 18.2 (1987). 5–13.
Sherwin, Adam. ' "Naked Dance" Advert Was a Cover-Up.' *The Times* (25 June 2003) sec. Home news: 6.
Shrum Jnr, Wesley Monroe. *Fringe and Fortune: The Role of Critics in High and Popular Culture*. (Princeton, NJ: Princeton University Press, 1996).
SIBMAS. *International Association of Libraries and Museums of the Performing Arts*. Available: www.sibmas.org/English/sibmas.html. Accessed: September 2005.
Siegel, Marcia B. *At the Vanishing Point: A Critic Looks at Dance*. (New York: Saturday Review Press, 1972).
——. *The Tail of the Dragon: New Dance 1976–82*. (Durham: Duke University Press, 1991).
——. *Watching the Dance Go By*. (Boston: Houghton Mifflin Company, 1977).
Snowdon, A. *Snowdon on Stage*. (London: Pavilion, 1996).
Sontag, Susan. *Against Interpretation*. (London: Eyre and Spottiswoode, 1967).
——. 'Film and Theatre.' *The Tulane Drama Review* 11.1 (1966). 24–37.
——. *On Photography*. (New York: Penguin Books, 1979).
Sparshott, Francis. 'Aesthetics of Music: Limits and Grounds.' *What Is Music?* Ed. Philip Alperson. (Philadelphia: Pennsylvania State University Press, 1987).
Steedman, Carolyn. 'The Space of Memory.' *History of the Human Sciences* 11.4 (1998). 65–83.
Steiner, George. *Real Presences*. (Chicago: University of Chicago Press, 1989).
Stewart, David, and Algis Mickunas. *Exploring Phenomenology*. (Chicago: American Literary Association, 1974).
Stewart, Gabe. 'Sounds, Smells and Style. But Little Substance.' Review. *Edinburgh Evening News* (1 September 1999) 24.
Stoppard, Tom. *The Real Inspector Hound and Other Entertainments*. (London: Faber and Faber, 1993).
Theatre Museum. Available: www.theatremuseum.org. Accessed: September 2005.
Thomson, Peter, Ed. 'Practice as Research (Notes and Queries).' *Studies in Theatre and Performance* 22.3 (2003). 159–80.
Törnquist, Egil. *Transposing Drama: Studies in Representation*. (London: Macmillan, 1991).
Tynan, Kenneth. *A View of the English Stage 1944–65*. (London: Methuen, 1984).
Van Camp, Julie. 'Copyright of Choreographic Works.' *1994–5 Entertainment, Publishing and the Arts Handbook*. Eds Stephen F. Breimer, Robert Thorne and John David Viera. (New York: Clark, Boardman, and Callaghan, 1994).
——. 'Review: Dance Criticism by Arlene Croce, Edwin Denby and Marcia Siegel.' *Dance Research Journal* 24.2 (1992). 41–4.
van Manen, Bertien. *East Wind, West Wind*. (Amsterdam: De Verbeelding, 2001).
——. *A Hundred Summers, a Hundred Winters*. (Amsterdam: De Verbeelding, 1994).
Varney, Denise, and Rachel Fensham. 'More-and-Less-Than: Liveness, Video Recording and the Future of Performance.' *New Theatre Quarterly* 16.1 (2000). 88–96.
Velody, Irving. 'The Archive and the Human Sciences: Notes Towards a Theory of the Archive.' *History of the Human Sciences* 11.4 (1998). 1–16.
Villeneuve, Rodrigues. 'Photography of Theatre: Images Always Fail.' *Canadian Theatre Review* 64. Fall (1990). 32–7.

Watson, Ian. ' "Reading" the Actor: Performance, Presence and the Synaesthetic.' *New Theatre Quarterly* 11.42 (1995). 132–46.

Williams, Christopher, Ed. *Realism and the Cinema*. (London: Routledge and Kegan Paul, 1980).

Williams, Raymond. *Television: Technology and Cultural Form*. (London: Fontana, 1974).

Wilson, Sue. 'Tales of Passion, Obsession and Tragic Isolation.' Review. *The Independent* (4 September 1999). Features, 8.

Wood, Julian. 'Repeatable Pleasures: Notes on Young People's Use of Video.' *Reading Audiences: Young People and the Media*. Ed. David Buckingham. (Manchester: Manchester University Press, 1993).

Wurtzler, Steve. ' "She Sang Live, but the Microphone Was Turned Off": The Live, the Recorded and the Subject of Representation.' *Sound Theory Sound Practice*. Ed. Rick Altman. (London: Routledge, 1992). 87–103.

Index

The page numbers in bold indicate figures.

Abercrombie, Nicholas, 88
academy, the, 41–8
  and archives, 41–3
  and authority, 5, 41, 44–5, 46, 48, 63–8
  and the canon, 47–8, 65–6
  and original research, 33, 41, 42, 242n
  and teaching, 2, 48, 56
acting (and actors), 18, 94, 97, 218–19
Adventures in Motion Pictures, *see* Bourne, Matthew
Advertising Standards Authority, 152, 154
Against Interpretation, 196–7, 200–3, 212, 235
Allen, Dave, 90, 101
Aloff, Mindy, 38
Alvarado, Manuel, 143, 144
American Film Theatre, 95–100, 108, 245n
American Theater Lab, 215
Anderson, Laurie, 10
Anderson, Lea, 170, 175, *see also* Featherstonehaughs, The
Arcade, Penny, 123, **124**
archive fever, 38–40
archives, 31–40, 53–4, 76
  and authority, 5, 31, 33, 34, 41, 53
  and fragmentation, 32, 36, 38, 43, 52
  and memory, 31, 41, 49–50, 52
  performing arts archives, 33–6
  and photography, 117, 118, 121
  promise of archives, 31–2, 42, 94, 99
  and research, 33, 41–2
  and video, 75–9, 82–3, 90, 94, 99
Artaud, Antonin, 11
artistic intention, 193–4, 249n
Audiences,
  contract with performance, 19
  expectations, 92, 95, 100, 101–3, 105, 109, 117, 147, 150, 152, 155
  and memory, 49, 51, 52
  see also being there live
aura, 24–5, 109–10, 246n
Auslander, Philip, 14–15, 25, 46, 79, 104, 109, 183, 240n, 241n, 243n
authenticity,
  archival, 41
  and photography, 115, 116, 132, 139, 164, 172, 179
  and video, 77–8, 82
  see also originary
authority
  of academy, 5, 41, 44–5, 46, 48, 63–8
  of archives, 5, 31, 33, 41
  of artists, 5, 67
  of photography, 115, 121, 123, 127, 130, 141, 143–4, 149
  of video, 50, 73, 77, 80, 82
  of writing (and reviews), 44, 184, 190–2, 195

Balanchine, George, 21
Banes, Sally, 197
Barba, Eugenio, 1, 10, 21, 51, 222, 237, 239n, 240n
Barnes, Clive, 189–90, 192
Barthes, Roland, 115, 132, 201–2, 241n
Baudrillard, Jean, 26, 93, 109
Bazin, André, 115, 245n
Beckerman, Bernard, 10, 18–19, 210–11, 215, 217, 249n
being there live, 2, 39–40, 102, 130, 135, 196, 232, 236–7, 238
Beloff, Hella, 180
Benjamin, Walter, 19, 24–5, 180, 246n
Bentley, Eric, 206
Berger, John, 146, 152, 234
betrayal, and documentation, 13, 24, 73, 204, 235
Billington, Michael, 187, 188, 218, 219, 248n, 250n

Blau, Herbert, 12, 217
Blind Faith (Yolande Snaith Theatre Dance), 172, **173**, 176–7
body, the (and bodies), 18, 123, 126, 199, 215, 216–18, 219–20, 221, 223
books, book form, 43, 44, 45, 49, 58, 63–5, 249n
Booth, John, 190
Bourne, Matthew, 176
Bradley, Hilary, 42
Brandon, Graham, 114, 117, 118, **119**, 121, 122, 144, 148, 246n
Breslauer, Jan, 147
Brith Gof, 44
British Sound Library, 65, 83, 240n
Broadway Theatre Archive, 50
Brook, Peter, 11, 49, 187, 217
Brooks, Bonnie, 21
Brooks, Virginia, 90, 93
Brown, Richard Harvey, 188
Bruce, Keith, 189, 191
Burgin, Victor, 146–7
Burnshaw, Stanley, 224
Burton, Richard, 94

canon, the, 41, 47–8, 65–6, 191, 243n
Carlson, Marvin, 51
Carr, C, 126
Cass, Joan, 192
CCTV, 91, 244n
CD/CD-Rom, 60, 61–2, 67
Chamberlain, Franc, 233, 242n
Channel 4, 88, 102, 103, 105
Christiansen, Rupert, 185, 187, 194, 207
Clanjamfrie, 44
Clifton, Thomas, 221, 222–3
closeness,
    and photography, 118, 120, 128
    and video, 85, 86, 102
Clurman, Harold, 188
collaboration, in documentation, 113, 122, 135, 136, 163, 169, 174–6
Companyia Carles Santos, 53
Complicite, 118, **119**
Connor, Steven, 110
Cooper, Neil, 185, 199, 207
Copeland, Roger, 46, 109, 195

copy, 16, 18, 19, 26, 27, 91–2, 93, 109, 120, 233
copyright, 241n
Cousin, Geraldine, 43–4
Crickmay, Anthony, 168–9
criticism,
    function of, 191–2, 199–204
    see also reviews
Croce, Arlene, 188, 196
Cubitt, Sean, 76, 78, 104
Cunningham, Merce, 10, 220
Cutler, Anna, 44

Daguerre, Louis, 115
Damaged Goods, 53,
    appetite, 166
    Visitors Only, **168**
dance,
    and anti-interpretative bias, 197, 202, 212
    disappearance of, 21, 22, 35, 80, 241n
    see also video dance
Dance Heritage Coalition, 22
Daniel, Claire, 78, 83–4
Davis, Jill, 47, 66
death, 10, 23, 38, 39
de Certeau, Michel, 241n
decisive moment (also privileged moment, perfect moment), 120, 137, 140, 212, 213, 233–4, 247n
deep focus cinematography, 98, 244n
definitive performances, 97, 218–19, 220
de Frutos, Javier, 175
Deleuze, Gilles, 19
de Marinis, Marco, 24, 91
Denby, Edwin, 196, 219–20
Derrida, Jacques, 25, 38–40, 49, 237, 241n
digital technology,
    and photography, xi, 170, 172, 174, 177, 180, 235, 246n
    and video, 75, 78, 79, 93, 104
disappearance,
    as radical ideology, 1, 12–13, 16–17, 26, 231, 237
    see also Phelan, transience

documentation,
  as driven by technology, 25–6,
  and the economics of production, 48, 55, 58–9, 65–6, 68–9, 148–9, 154–6
  and the future, 23, 35, 65, 79, 83, 97, 115
  and gaining wider audiences, 58, 80, 85, 95, 96, 100, 102, 130, 244n
  as saving performance, 21–3, 34–6, 43–4, 45–6, 80–1, 93, 97, 114, 195, 233
  as speaking for performance, 3–4, 8, 27, 39–40, 41, 68–9, 233–6, 238
Dodds, Sherril, 83, 101–2, 108, 245n
Durant, Alan, 92
DV8, 48, 102–7, 245n
  Cost of Living, The, 102, 103, 104, 105–7
  Dead Dreams of Monochrome Men, 102
  Enter Achilles, 102, 103
  Strange Fish, 102

Early, Fergus, 168–9
Edgerton, Harold, 141
Elam, Keir, 45
electronic theatre, 94–5
Elsom, John, 190
Elwers, Catherine, 14, 77–8, 245n
embodiment,
  embodied knowledge, 138, 215–16, 221, 250n
  and writing, 199, 209, 215–17, 219, 222–5, 228–9
  see also body, the
ephemerality, *see* transience
Erenstein, Robert, 48
Etchells, Tim, 57, 59, 60, 63, 64–5, 67, 84, 85, 86, 160, 174, 176, *see also* Forced Entertainment
Evans, Martyn, 88
Evans, Richard, 33
Ewing, William, 139

Featherstonehaughs, The
  Immaculate Conception, 177, **178**

The Featherstonehaughs Draw on the Sketchbooks of Egon Schiele, 170, **171**
Feingold, Michael, 50
Fensham, Rachel, 45, 46, 50, 52, 83
Féral, Josette, 11, 190, 203
Feuer, Jane, 75, 244n
Finley, Karen, 123, **125**, 126–7, 174
Fish, Stanley, 249n
Fisher, Mark, 186, 191, 193, 209
Forced Entertainment, 5, 53, 55–69, 84–7, 157–62
  and artistic proliferations, 59–63, 68, 160–2
  Club of No Regrets, 85, 160, **161**, 162, 174
  Emanuelle Enchanted, 60–1, 63, 66, 158, **159**
  First Night, 176
  Imaginary Evidence, 60, 61–2, 67–8
  Marian & Lee, 84–5
  Nights in This City, 63–4
  and photography, 157–62
  and pragmatic documentation, 57–9, 68, 85–6, 157–8
  reputation, 58–9, 65, 68–9
  Some Confusions in the Law About Love, 86, 89
  200% and Bloody Thirsty, 85
  and video, 56, 57, 84–7
fragmentation, *see* archives, postmodernism
framing,
  and photography, 117–18, 121, 122, 128, 139–40, 158, 160, 162, 164, 172
  and video, 98–9, 103
Freebalm, Alison, 186, 190, 199
Freshwater, Helen, 42, 46, 237
Fuchs, Elinor, 88
functional analysis, 203

Garner, Stanton, 216
Genet, Jean, 20–1
geno-text, 202
Giesekam, Greg, 44–5
Glendinning, Hugo, 60, 114, 157–8, **159**, 160, **161**, 162, 163, 174, 176, 247–8n

Goldberg, RoseLee, 13–14, 126
Goldsworthy, Andy, 242n
Greenfield, Lois, 114, 136–7, **138**, 139, **140**, 141, **142**, 144, 163, 248n
Guest, Ann Hutchinson, 22, 241n

Hall, Fernau, 22
Hall, Peter, 1, 98, 100, *see also Homecoming, The*
Harper, Douglas, 234
Heathfield, Adrian, 12, 13, 66, 238
Helmer, Judith, 64, 67
Henry IV (The Old Vic), 218
Het Zuidelijk Toneel, *see India Song*
Homecoming, The (Royal Shakespeare Company), 96, 98–100, 245n
Husserl, Edmund, 216
hybrid performance, 5, 91, 97, 100–1, 105–7

Iball, Helen, 47–8
Imhauser, Marcelle, 108
India Song (Het Zuidelijk Toneel), 163–4, **165**, 166, **167**, 168, 185–8, 190–1, 193–4, 197–9, 206–9
ineffability, 18, 46, 238, 250n
internet, 48, 58, 243n
intersubjectivity, 216–17, 218, 224, 227–8

Jacobson, Colin, 123
Jarrell, Randall, 200–1
Jones, Amelia, 2, 237
Jonson, Ben, 21–2
Jowitt, Deborah, 193, 195, 197

Kafka, Franz, 250
Kalb, Jonathan, 187, 194, 248n
Kaye, Nick, 61, 63–4, 66
Keller, Hans, 192, 203
kinesics (and kinaesthesia), 198, 215
Kirby, Michael, 9, 24, 61, 80, 197, 200
Kirkley, Richard, 94–5
Koestler, Arthur, 224

Landau, Ely, 96, 97
Langer, Susan, 11
liveness, 8, 14–15, 18, 46, 73–4, 87–8, 94, 104–5, 109–10
Lockyer, Bob, 81

Longhurst, Brian, 88
lost performances, 22, 76–7, 79
Lumière, Auguste, 115

Macaulay, Alastair, 186, 187, 190
Malzacher, Florian, 64, 67
Martha Graham Dance Company, **138**
mass production, 15, 16, 19
McAdams, Dona Ann, 114, 122–3, **124**, **125**, 126–8, **129**, 130, 144, 163, 174
McAuley, Gay, 9, 23–4, 26, 37, 76, 81, 90, 235
McCausland, Elizabeth, 248n
McMillan, Joyce, 186, 188, 191, 194, 198, 208
Melrose, Susan, 192, 250n
Melzer, Annabelle, 24, 81–2, 108
memory, 2, 5, 12, 20, 26, 31, 38, 48–53, 188, 218, 229, 238
  and archives, 31, 38, 41, 48–50, 52, 241n
  and audiences, 2, 3, 9, 48, 49, 51–2, 54, 65, 120, 239n, 240n
  and photography, 120, 122, 126, 152
  transformational power, 51, 53
  and video, 50, 56, 57
  and writing, 50, 188, 205, 218
Merleau-Ponty, Maurice, 215, 221
Miller, Jonathan, 19, 99, 233
Mitchell, Jack, 247n
Mousetrap, The, 16
music,
  criticism, 192, 201–3, 211–12, 222–4, 250n
  notation, 22, 241n
  recorded music, 27, 92
musicals, 16–17, 20, 244n
Myles, Euan, 114, 149–50, **151**, 152, **153**, 155, **156**, 163

Nachtasiel (*The Lower Depths*, Ro Theatre), 130, **131**, **133**, 134, 135
Nash, Chris, xi–xii, 114, 169–70, **171**, 172, **173**, 174–7, **178**
National Endowment for the Arts, 127, 241n
National Initiative to Preserve America's Dance, 241n

National Review of Live Art, 56
Newson, Lloyd, 102, *see also* DV8
New York Public Library, 34, 35

objectivity (and neutrality), 31, 32, 50, 61, 80, 82, 86, 91, 118, 126, 183, 235, 246n
originary (and origins, original), 12, 16, 19, 26, 33, 36–7, 39–40, 49, 67, 73, 81, 92, 96, 100, 110, 177
Otake, Eiko, 88–9

Pavis, Patrice, 12, 24, 49, 187, 201, 213, 232
Pavlova, Anna, 115
Peña, Richard, 96, 97
performance art (and live art), 2, 10, 13–14, 18, 44–5, 66, 74, 122, 123, 126, 197, 250n
performance studies, 2, 8, 9, 14, 36, 42, 46, 56, 236
performative now, 10–11, 12, 17, 18, 19, 88, 104, 210, 211, 232, 240n
performative writing, 184, 205–6, 207, 208–9, 226–30, *see also* reviews, writing
personal video recorder (PVR), 79, 104–5
Phelan, Peggy, 1, 12–13, 14, 19, 20, 25, 26, 36, 58, 211
phenomenology, 216, 221, 234
Phillips, Sïan, 240n
photo-call, 118, 122, 136, 246n
Photo-choreography, 127
photography, 6, 113–45, 146–80, 210, 233–4, 242n, 246–8n
and anchoring text, 132, 135, 150, 158
and archives, 117, 118, 121, 148
and authority, 115, 121, 123, 127, 130, 141, 143–3, 149
during live performances, 122
as history, 115, 117, 122, 147, 247n
and manipulation, xi, 116, 139, 152, 170, 172, 173, 177, 179, 180
and movement, 137–8, 164, 170, 235

narratives of production, 114, 123, 136, 139, 141–2, 143–3, 160, 162, 179
pre-publicity photographs, 149, 154, 157, 174–7, 247n
publicity photographs, 6, 114, 146–9, 154–6, 247n
reproducibility, 127, 130
and responsibility, 148, 152
revelation, 6, 113, 114–16, 121, 136–8, 139–41, 144–5, 170, 172, 177, 180, 246n
studio, 122, 136, 140, 152, 170
and transformation, xi–xii, 114, 116, 137, 143–145, 152, 163
see also closeness, digital technology, framing, memory, photojournalism, witness
photojournalism, 123, 130, 132, 134, 135, 139, 246n
Pollock, Della, 184, 205–6, 226
Porter, Andrew, 223–4, 250n
postmodernism, 14, 109–10
and fragmentation, 2, 32, 52, 54, 60–3, 66–7, 160
and history (and archives), 32–3, 52, 54, 241n
and photography, 160, 180
and video dance, 101–2, 108
power, *see* authority
practice-based research, 3, 43, 242n, 243n
presence, 17, 18–19, 37, 46, 87–8, 104–5, 106, 123, 126, 128, 215, 217–20, 222–3, 227–9
production run, 15–19, 20, 51, 240n
PS122, 114, 122, **124**

Quick, Andrew, 13

realism,
and photography, 132, 134, 135, 137, 246n
and video (and film, television), 95, 101, 106, 108, 244n, 245n
Rebellato, Dan, 16
repetition, 11, 16, 17, 19–20, 26, 79, 231, 240n

reviews, 6–7, 183–99, 205–10, 212–14, 215–21
  and authority, 184, 190–2, 195, 248n
  and description, 187, 194, 195–9, 200–1, 206
  and evaluation, 189–92, 193–4, 199, 207–8
  as history, 184–5, 195
  and interpretation, 192–5, 196–7, 200–4
  'knocking copy', 187–8, 207–8, 248n
  narratives of production, 186–9
  overnight review, 186, 187, 188, 189
  see also criticism, memory, writing
Rich, Frank, 196
Rigg, Diana, 248n
Ro Theatre, see Nachtasiel
Rogoff, Gordon, 187
Rosenberg, Douglas, 79, 82, 92, 101, 240n
Rosenthanl, Daniel, 94
Royal Court Theatre, **118**, 250n
Royal Shakespeare Company, 98, 219, 240
Rush, Michael, 245n
Russell, Susan, 17
Rye, Caroline, 46, 243n

Sartre, Jean-Paul, 217
Savedoff, Barbara, 116
Schechter, Joel, 179
Schneider, Rebecca, 9, 37, 242n
Senelick, Laurence, 26, 114–15
Sher, Antony, 240n
Shrum, Wesley Monroe, 23, 191
SIBMAS, 42
Siegel, Marcia, 9, 24, 50, 195, 212–13, 215–16, 220
Snowdon, A, 128
Snyder, Allegra Fuller, 79
Sokolow, Anna, 220
Sontag, Susan, 77, 115, 116, 120, 137, 196–7, 200
Sparshott, Francis, 215
Sprinkle, Annie, 128, **129**
Steedman, Carolyn, 32, 33, 49, 52
Steiner, George, 201, 202

Stewart, Gabe, 188, 208
Stuart, Meg, see Damaged Goods
synaesthesia, 208, 221–5, 228, 250n

technological contingency, 75–6
television, 75, 77, 19, 87–90, 91, 95, 101, 104–9, 120, 244n, 245–6n
  broadcast flow, 79, 103, 104–5, 245–6n
  watching television, 87–90, 103, 104–8, 244n
Theatre Museum (Victoria and Albert Museum), 34, 35, 37, 83, 117
theatrofilm, 93, 94
Törnquist, Egil, 233
transience,
  and authenticity, 11, 39–40
  fear of, 8, 21–3, 35–6, 43, 44, 79, 115, 231
  of live performance, 1–2, 8–13, 15–16, 20, 22, 24, 26–7, 39, 43–5, 48, 49, 51, 54, 56, 59, 67, 195, 210, 231, 237, 240n
  positive valuation of, 8, 11, 13, 20–1, 23–4, 53–4, 73–4, 98
  of video, 78–80
Traverse Theatre Company, 149–50, 155–6
  Olga, 152
  Outlying Islands, 152, **153**
  When the Bulbul Stopped Singing, 150, **151**
Trisha Brown Company, 212
truth, ideas of, 2–3, 50, 52, 86, 118, 121, 143–4, 154, 180
Tynan, Kenneth, 195, 196, 218–19

Ultima Vez,
  Scratching the Inner Fields, 53

Van Camp, Julie, 196
Vandekeybus, Wim, see Ultima Vez
Van der Burght, Chris, 114, 163–5, **165**, 166, **167**, **168**
van Hove, Ivo, 163, 185, 187, 190, 194, 198
van Manen, Bertien, 114, 130, **131**, 131, **133**, 134–6, 144
Varney, Denise, 45, 46, 50, 52, 83

Velody, Irving, 33, 49
video, 5–6, 58, 73, 91, 104–5, 233, 243n
  accessibility of, 77–8, 80, 95
  and adaptation, 74, 81–2, 102, 244n
  and archives, 78–9, 82–3, 90, 94, 117
  and authenticity, 77–8, 86
  and authority, 50, 73, 77, 80, 82, 87, 90
  as document, 74, 76–8, 80–4
  ephemerality of, 78–80
  and liveness, 73–7
  and television, 75, 77, 88, 90, 91
  as a verb, 75–6, 77, 84
  watching video, 74, 76, 78, 82, 84–7, 88–9, 90–1, 100, 104–5, 107–10
  see also CCTV, closeness, digital technology, framing, video art, video culture
video art, 77–8, 79, 245n
video culture, 76, 87–9, 91, 107, 117
video dance, 92, 100–2, 103, 106–7, 108–9, 110, 139, 245n
Villeneuve, Rodrigues, 2, 122, 147

Watson, Ian, 215
Wilder, Thorton, 10
Williams, Raymond, 104
Wilson, Sue, 186, 193
witness,
  audiences as, 49, 51, 85, 126, 217, 228
  documentation as, 43
  photography as 121–2, 126, 127–8, 210
Wood, Julian, 87
writing,
  and embodiedment, 199, 209, 215–17, 219, 222–5, 228–9
  and space, 214, 216, 227, 228–9
  and time, 210–14, 227, 229
  see also criticism, performance writing, reviews
Wurtzler, Steve, 109, 246n

Yolande Snaith, 176–7, *see also Blind Faith*

Zandwijk, Alize, 130, 132, 134
Zimmer, Elizabeth, 73, 93